Introduction to
IMPACT ENGINEERING

Introduction to
IMPACT ENGINEERING

M. Macaulay

Senior Research Fellow
Brunel University

London   New York
CHAPMAN AND HALL

First published in 1987 by
Chapman and Hall Ltd
11 New Fetter Lane, London EC4P 4EE
Published in the USA by
Chapman and Hall
29 West 35th Street, New York NY 10001

Printed in Great Britain by J.W. Arrowsmith Ltd, Bristol

ISBN 0 412 28930 X

British Library Cataloguing in Publication Data

Macaulay, M.
   Introduction to impact engineering.
   1. Impact   2. Strength of materials
   I. Title
   620.1'125   TA418.34

ISBN 0–412–28930–X

Library of Congress Cataloging in Publication Data

Macaulay, M., 1926–
   Introduction to impact engineering.

   Bibliography: p.
   Includes index.
   1. Impact. I. Title. II. Title: Impact
engineering.
TA354.M33   1986   620.1'125   86–12983
ISBN 0–412–28930–X

Contents

Preface

We are all familiar with impact. Lesser impacts such as hammering a nail, cracking an egg or stubbing a toe are part of everyday life. More violent impacts such as those caused by car crashes or bullets are fortunately less common but are still well enough known to be taken for granted. Very violent impacts such as meteorites striking the earth are outside our personal experience but we are aware of them.

Despite this, impacts remain mysterious. They occur too quickly for us to follow what is happening and the evidence they leave behind is often ambiguous. Over the last thirty years improvements in high speed instrumentation and developments in computing have made them more comprehensible and an increasing amount of attention is being paid to the subject which is an area of expanding scientific and engineering research.

A multi-disciplinary approach is not yet established and information is scattered in many places and expressed in a variety of jargons. In applied mathematics, impacts have provided interesting theoretical problems with elegant solutions but it has been difficult to check results experimentally. Impacts can change the behaviour of materials but similar changes can sometimes be produced in other ways and the underlying mechanisms are not clear. Empirical solutions to engineering problems have worked reasonably well but it is hard to know what to do if things go wrong.

The aim of this book is to provide an introductory guide at a level which is intelligible to the non-specialist – but in sufficient detail for an interested reader to move on to specialist published work. The material presented has been selected on the basis of what, in the light of my own experience, has been found to be particularly useful.

1
Linear elasticity

1.1 INTRODUCTION

1.1.1 General

(a) Scope of chapter

Analysis of the behaviour of materials under load is carried out on mathematical models of idealized materials which approximate to real materials but behave more simply and consistently. Usually a model can be constructed to give sufficiently accurate results over a limited range of behaviour, but a single model is unlikely to give adequate results over all the ranges of behaviour of interest.

The simplest case to model is linearly elastic behaviour. This covers many cases of importance in static loading and, though it is rather less applicable to impact loading because impact loads tend to be too high for linearly elastic behaviour to occur, it gives a firm and well-documented basis from which to study more complex behaviour.

Rigorous analysis of loads and deformations in three dimensions is complex and sometimes is either impossible, using current techniques, or gives rise to such complexity that the results are difficult to understand and apply. Much simpler analysis is often used to give a qualitative understanding and to help classify and rationalize the behaviour of real materials.

The aim of this chapter is to describe, as simply as possible, the behaviour of an idealized, linearly elastic material under impact loading and to relate this to the behaviour of real materials whilst they are behaving in a manner which is approximately linearly elastic.

(b) Eulerian and Lagrangian co-ordinates

The behaviour of a moving body needs to be related to a system of reference co-ordinates. Two systems are in common use. In the Eulerian system the reference co-ordinates are fixed in space and the body moves relative to them. This is useful when displacements are large but it can be difficult to deal with the behaviour of the surface of the body.

In the Lagrangian system the moving body and the reference co-ordinates are both regarded as being stationary with the rest of the environment moving.

This is useful if the behaviour of the surface of the body is being considered but it can be difficult to deal with large displacements (Zukas *et al.*, 1982, pp. 372–7).

(c) Condensed notation

Analysis in three dimensions inevitably involves a large number of variables. This bulk of information makes the analysis very tedious and can lead to important factors being obscured by a mass of detail. Much of this detail is repetitive, because identical types of expression are needed for each of the three dimensions, and this is made use of in condensing the notation and the analysis. Condensed analysis involves the manipulation of tensors and matrices and is beyond the scope of this book, but the condensed notation appears widely in the literature and will be used where appropriate (Ford and Alexander, 1977, pp. 96–105).

1.1.2 Basic relationships

(a) Strain rate

Strain ε is a dimensionless parameter which describes the amount of deformation which occurs in a piece of material when it is loaded. The rate at which deformation occurs under impact loading is expressed as the strain rate $\dot{\varepsilon}$, which is generally given in units of strain per second (s^{-1}). For a given total strain there is an interrelationship between strain rate and the duration of the straining process; high strain rates occur over short time scales and vice versa. Generally it is only the order of magnitude of a strain rate which is important and minor changes in strain rate can be ignored. Some strain rates of practical interest are shown in Fig. 1.1.

(b) Wave propagation

Deformation does not occur instantaneously throughout a material but propagates at a finite rate, moving through the material in a wave. As with all waves, reflection, refraction, dispersion and interference occur but these effects are ignored in Chapters 1–4 and only two limiting conditions are dealt with. In one of these wave propagation is ignored and deformations are assumed to occur simultaneously throughout the material. Because wave speeds are often very high in the context of the time scales being considered, this assumption gives accurate results in many cases.

In the second case, propagation of a single, simple wave is considered. It is assumed that there is a sudden discontinuity at the wave front so that the strain rate there is infinite in theory and, in practice, is determined by the very short response time of individual atoms. Behind the wave front the nominal strain rate is taken as the strain divided by the time since the wave front passed.

For a homogeneous, isotropic, linearly elastic solid a wave of deformation

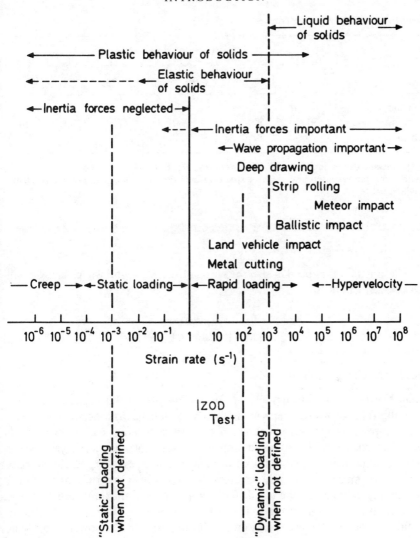

Figure 1.1 Range of strain rates found.

travels at a constant speed which depends on the stiffness and density of the material and the type of deformation involved. It does not depend on the magnitude of the deformation. Table 1.1 gives elastic wave speeds for some common materials under standard conditions.

Small deformations in a linearly elastic fluid also have a constant wave speed but with large deformations the wave speed varies with the magnitude of the deformation.

Table 1.1 Elastic wave speeds.

	Elastic wave speeds (ms^{-1})					
	Steel	*Aluminium*	*Cast iron*	*Copper*	*Lead*	*Glass*
Tensile and						
compressive	5000	5000	3900	3650	1200	5250
Shear	3200	3050	2450	2250	700	3200

These are speeds for simple stress waves given by $C = \sqrt{E/\rho}$ and $C_T = \sqrt{G/\rho}$.
With more complex stress systems wave speed depends also on Poisson's ratio v and is usually slightly higher.

The elastic wave speed is also the speed of sound, sound waves being small elastic disturbances.

1.2 SIMPLE ANALYSIS

1.2.1 The elastic limit of solids

Behaviour remains linearly elastic for an idealized solid material up to a limiting value of stress, after which other types of behaviour occur. The elastic limit is specified by the stress at which it occurs, called the *yield stress* σ_y. With an idealized material the elastic limit is clear-cut and the tensile yield stress $+\sigma_y$ is numerically equal to the compressive yield stress $-\sigma_y$. Real materials differ in various ways from the idealized one, usually in that the elastic limit is not clearly defined and the tensile and compressive elastic limits are different.

At high rates of strain the yield stress can be two or three times the static value, the increase tending to be greater for materials with a lower static yield stress. This is shown in Fig. 1.2(a) (see Goldsmith, 1960, Fig. 280). For many materials showing a strain rate effect the yield stress varies approximately linearly with the log of the strain rate up to rates of about 10^3 s^{-1}, as shown in Fig. 1.2(b) (see Campbell, 1970, Fig. 59).

Yield stress can also vary with temperature, increasing with decrease in temperature as shown in Fig. 1.2(c) (see McClintock and Argon, 1966, Fig. 17.3; Campbell, 1970, Fig. 12). In general a reduction in temperature has the same qualitative effect as an increase in strain rate. This is discussed further in Chapter 2.

Static yield stresses for some common materials are given in Table 1.2.

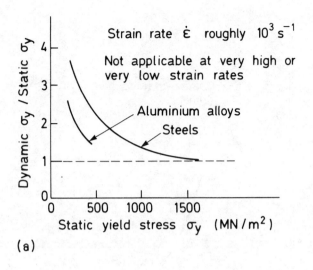

(a)

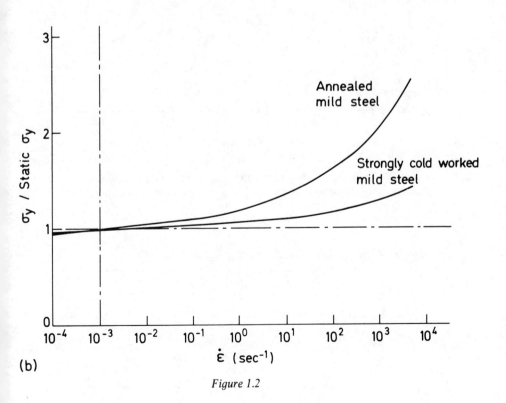

(b)

Figure 1.2

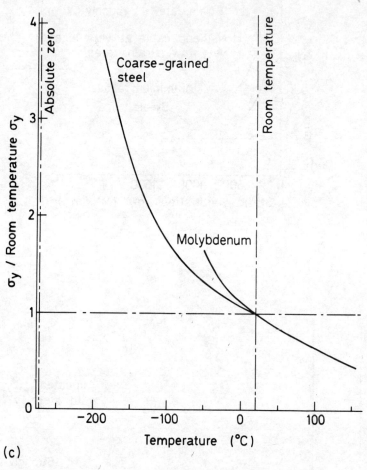

(c)

Figure 1.2 (a) Increase in dynamic yield stress at room temperature. (b) Variation in σ_y with $\dot{\varepsilon}$ at room temperature. (c) Variation in σ_y with temperature under static loading.

Table 1.2 Typical static yield stresses (MNm^{-3}).

	Static yield stress (MNm^{-3})			
Steel	Aluminium	Cast iron	Copper	Lead
300 to 800	250	150 tension 700 compression	200 to 350	10

1.2.2 Propagation of a stress wave

Although either stress or strain could be taken as a reference parameter, deformation waves in solids are generally referred to as *stress waves*. In many cases stress and strain occur simultaneously, but often strain lags behind stress; consequently, stress makes a more convenient reference.

Elastic tension or compression produces a change of volume but no angular distortion, so waves of direct stress are also called *dilatational* or *irrotational waves* (Kolsky, 1963, p. 13).

It is simple to study the propagation of a direct stress wave in one dimension, using the concepts of impulse and momentum (see Chapter 5). A compressive wave is considered but a tensile wave is identical apart from the changed sign of the stress.

The momentum of a body of constant mass M moving in a straight line with constant velocity v is Mv. The impulse produced by a constant force F acting for a time t is Ft. During an impact the change in momentum which occurs is equal to the impulse applied, giving

$$M(v_2 - v_1) = Ft$$

where v_1 is the initial velocity of the mass and v_2 is the final velocity.

If a straight rod of linearly elastic material travelling at velocity v hits a stationary, rigid, flat surface perpendicular to its direction of travel, the end of the rod stops abruptly and a compressive stress wave of amplitude $-\sigma$ propagates along the rod at wave velocity C. Using Eulerian co-ordinates with the origin at the point of contact gives the following conditions at time t after the initial contact of the rod with the surface (see Fig. 1.3).

The original length of the stationary part of the rod l is Ct. The reduction in

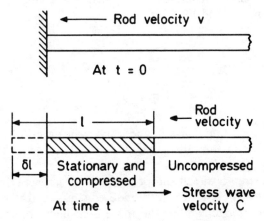

Figure 1.3 Linearly elastic rod striking rigid flat surface.

length l is δl; this is the distance that the end of the rod would have travelled if it had not been stopped, i.e. vt. Hence

$$\delta l = vt$$

The compressive strain in the stationary part of rod is given by

$$-\varepsilon = -\frac{\delta l}{l}$$

Hence

$$-\varepsilon = -\frac{v}{C} \tag{1.1}$$

The change in momentum is given by

$$M(0 - v) = -Mv$$

and the mass by

$$M = \rho_0 ACt$$

where ρ_0 is the density of the unstressed rod and A is the cross-sectional area. The compressive force on the end of the rod is

$$F = -\sigma A$$

Now

$$\text{impulse} = \text{change in momentum}$$
$$-\sigma At = -\rho_0 ACtv$$
$$\sigma = \rho_0 Cv \tag{1.2}$$

So far the analysis applies for any relationship between stress and strain. To go further it is necessary to specify a relationship between stress and strain.

For a linearly elastic material, $\sigma = E\varepsilon$, where E is Young's modulus. Using this in Equations 1.1 and 1.2 gives

$$\frac{v}{C} = \frac{\rho_0 Cv}{E}$$

Hence

$$C = \pm\sqrt{\frac{E}{\rho_0}} \tag{1.3}$$

The wave velocity is thus constant and is independent of stress. The positive and negative solutions indicate that the wave could travel equally well in either direction along the rod.

The induced stress is directly proportional to the impact velocity and there is a limiting velocity above which the induced stress is greater than the yield stress σ_y. For short loading times the appropriate high strain rate value of σ_y applies but for long loading times the static value applies. Thus it depends on both impact speed and duration of impact whether the yield stress is exceeded or not.

The related situation in which a stationary rod is struck by a rigid moving mass is more conveniently dealt with using Lagrangian co-ordinates with the origin at the point of contact between the mass and the rod. In this case the rod is stationary ahead of the wave front and moving at velocity v behind it. The velocity of the moving part of the rod relative to the stationary part is called the *particle velocity*. The concept of particle velocity is very useful and is widely used.

With a wave of compression stress the particle velocity is in the same direction as the velocity of the wave front. With a wave of tensile stress it is in the opposite direction.

In experiments it has been found that induced stress, particle velocity and wave speed in rods are in broad agreement with the results of simple theory. There are discrepancies for short duration pulses which arise largely from neglecting the effects of secondary, transverse strains, and there are considerable experimental difficulties in obtaining a sufficiently accurate axial impact (Conway and Jakubowski, 1969).

The wave speed of a shear stress wave is $C_T = \pm \sqrt{(G/\rho)}$, which is lower than the wave speed of a direct stress wave because G, the shear modulus, is smaller than E.

In practice the induced stress and the shear wave velocity are found to agree well with theory, and there are no significant discrepancies because there are no secondary strain effects. On the other hand there are variations in the theoretical shear stress wave velocity with the shape of the cross-section.

The analysis of a stress wave in a body of elastic fluid with a constant cross-section of area A is basically the same as for an elastic rod. It is convenient to regard the fluid as being contained in an inextensible pipe. There is an initial static pressure P_0, and an increase to P. As the cross-sectional area remains constant, all of the reduction in volume must come from the longitudinal strain, giving the *volumetric strain* as $\bar{\varepsilon} = v/C$. Equating impulse and momentum, as before, gives

$$P - P_0 = \rho_0 C v$$

As $P - P_0 = K\bar{\varepsilon}$, where K is the bulk modulus, it follows that

$$C = \pm \sqrt{\frac{K}{\rho_0}} \tag{1.4}$$

Because $K = \rho(\mathrm{d}P/\mathrm{d}\rho)$, it follows that

$$C = \pm \sqrt{\left(\frac{\rho}{\rho_0}\frac{\mathrm{d}P}{\mathrm{d}\rho}\right)}$$

When changes in pressure are small, $\rho \simeq \rho_0$, so that

$$C = \pm \sqrt{\frac{\mathrm{d}P}{\mathrm{d}\rho}}$$

1.2.3 Graphical representation of a stress wave

In Fig. 1.4 the simple stress wave front is plotted in terms of distance travelled versus time. The straight line 0A represents the wave front given by

$$x = Ct$$

Note that Fig. 1.4 is not a graph of a function x which varies continuously with t. It represents a discontinuity, and it is only along the wave front 0A that a specific relationship between x and t exists. Elsewhere all that can be said is that region 1 represents that part of the rod which the wave front has not yet reached, and region 2 represents that part of the rod which the wave front has passed.

The plane defined by x and t is called the *characteristic plane* and the line 0A is called a *characteristic line* for the stress wave (Hopkins, 1968).

If the applied stress at the impact end varies with time as shown in Fig. 1.5, any requisite number of characteristic lines can be drawn, parallel to 0A, each representing the stress at a single instant of the stress pulse. Each stress propagates along its characteristic line $x = Ct$. The resulting variation in stress with time can then be plotted for any desired value of x. Because there is no change in the stress as it travels along, this variation is identical to the original variation at the impact end. Alternatively the variation of stress along the rod can be plotted for any desired value of t.

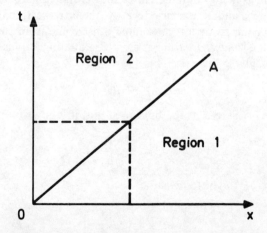

Figure 1.4 Characteristic line of a simple stress wave.

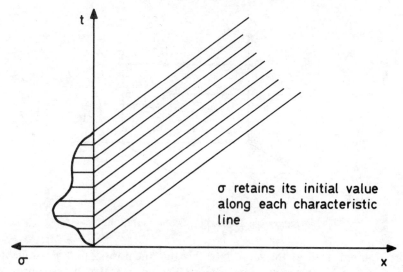

Figure 1.5 Characteristic lines for a stress pulse.

1.3 MORE COMPLEX ANALYSIS

Analysis is often carried out in much more detail than in the simple examples given above. In general, the behaviour of a very small element of the material is considered and this is then extended to the material as a whole. For linearly elastic behaviour the analysis is simplified considerably by assuming that the strains are sufficiently small for second-order effects to be neglected. The shape of the element is chosen to suit the problem in hand, common shapes being rectangular blocks, tetrahedrons, cylinders and spheres.

Under the action of applied stresses the element can move bodily, rotate and deform. It can do this slowly so that dynamic effects can be neglected, or quickly so that they need to be considered. If all possible effects were considered the analysis would be intractable; thus various approximations are made and various effects are neglected. These vary with the aspect being studied, and so results from different analyses may not be interchangeable and anomalies may occur.

1.3.1 Two- and three-dimensional behaviour

(a) Notation
Three-dimensional notation is discussed briefly using, as an example, the small rectangular block shown in Fig. 1.6. By convention, the orientation of any face of the block is denoted by the direction of the normal to it. Face ABCD has its

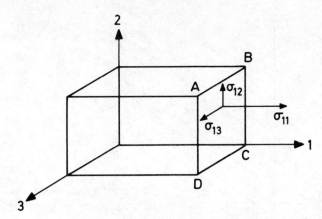

Figure 1.6 Three-dimensional notation.

direction indicated by the subscript 1, because the normal to it is parallel to axis 1. The positive direction of the axis is also taken as the positive direction for the normal. Any stress system acting on face ABCD can be resolved into a direct stress component parallel to axis 1 and two shear stress components parallel to axes 2 and 3. The direction of the stress component is given by a second subscript indicating the direction of the axis to which it is parallel. The direct stress component on face 1 is thus called σ_{11} and the two shear components are called σ_{12} and σ_{13}. The complete stress system on the face is given by $\sigma_{11} + \sigma_{12} + \sigma_{13}$. The axes can be represented by any appropriate symbols; a set of axes x, y, z would give the stress system as $\sigma_{xx} + \sigma_{xy} + \sigma_{xz}$.

The stress system on the face parallel to ABCD comprises the balancing set of stress components and is not specified separately. The complete stress system on the block is specified by the three stress components on each of the three faces whose normals are positive in the same directions as axes 1, 2 and 3. It can be represented by the matrix

$$\begin{vmatrix} \sigma_{11} & \sigma_{12} & \sigma_{13} \\ \sigma_{21} & \sigma_{22} & \sigma_{23} \\ \sigma_{31} & \sigma_{32} & \sigma_{33} \end{vmatrix}$$

which can be abbreviated to $[A_{ij}]$ $(i, j = 1, 2, 3)$ or even more simply to A. In indicial notation the complete stress system is

$$\sigma_{ij} \quad (i, j = 1, 2, 3)$$

The values of i and j, once defined, are not written again subsequently; so the indicial notation, for instance, would be simply σ_{ij}.

When the rectangular block lies at an angle to the axes its orientation needs to be taken into account. This is done for each face by stating the cosines of

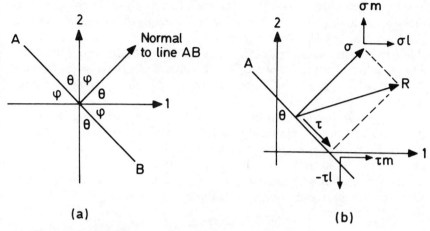

Figure 1.7 (a) Direction cosines. (b) Stress components in two dimensions.

the angles made between the normal to the face and each of the three axes. These are called *direction cosines*. Figure 1.7(a) illustrates the principle in two dimensions. Line AB is inclined at angle ϕ to axis 1 and angle θ to axis 2. The direction cosine of the normal is $\cos \theta$ with respect to axis 1 and $\cos \phi$ with respect to axis 2. Often $\cos \theta$ is called l and $\cos \phi$ is called m. Only one direction cosine needs to be known for two dimensions because θ and ϕ are complementary angles and the second cosine can be derived. Other relationships can be established too, in particular $l^2 + m^2 = 1$.

Figure 1.7(b) illustrates stress components in two dimensions. If a stress R acts on AB this can be resolved into a direct stress component σ normal to AB and a shear stress component τ parallel to AB. These can be further resolved into direct and shear stress components parallel to the axes. For the direction of R shown in Fig. 1.7(b) this gives σl and τm parallel to axis 1 and σm and $-\tau l$ parallel to axis 2. One of the shear components will always be in the opposite sense to the other three components, whether the direct stress is positive or negative, because a stress acting in either direction along AB will have one component in a positive direction and one in a negative direction.

The stress system on AB can be represented by the matrix

$$\begin{vmatrix} \sigma_{11}l & \sigma_{12}m \\ \sigma_{21}l & \sigma_{22}m \end{vmatrix}$$

In condensed notation this reduces to $[A_{ij}l_j]$ or $\sigma_{ij}l_j$. The direction cosines being referred to as l_1 and l_2 instead of l and m.

In three dimensions, three direction cosines are needed to define the orientation of a plane, although only two need be known because the third can be derived. The direction cosines for face 1 of the block with respect to axes 1, 2 and 3 are often called l_1, m_1 and n_1 and correspondingly for faces 2 and 3 they are l_2, m_2, n_2 and l_3, m_3, n_3. Relationships exist among the direction cosines; in particular, for any face, $l^2 + m^2 + n^2 = 1$.

The overall stress system for the complete block is obviously cumbersome to express in full, but the condensed notation remains exactly as for the two-dimensional system except that both i and j are considered up to the value 3 instead of only to the value 2.

With an isotropic material the condensed notation is very convenient, but properties of a linearly elastic composite material are not isotropic and can vary in three mutually perpendicular directions. It is often necessary to analyse such behaviour in detail, and it can be more convenient to introduce a contracted notation where stresses and strains are numbered in a pre-arranged sequence. A typical form of contracted notation would be

$$\sigma_1 = \sigma_{11}$$
$$\sigma_2 = \sigma_{22}$$
$$\sigma_3 = \sigma_{33}$$
$$\sigma_4 = \sigma_{12}$$
$$\sigma_5 = \sigma_{13} \qquad \text{etc.}$$

Such a notation will generally be specified as it arises.

When wave propagation effects are important, such a complex stress system gives rise to complex stress wave systems which can produce transient effects different from the steady-state principal stress patterns. There does not seem to be a developed stress wave theory analogous to the static principal stress theory.

Waves of direct stress propagate faster than waves of shear stress, and this needs to be reconciled with the results of static analysis which shows that direct and shear stresses are interlinked and occur together. Shear stress is equivalent to equal and opposite direct stresses at right angles to each other and at 45° to the direction of the shear stress, so that a wave of shear stress can be regarded as two waves of direct stress travelling at 45° to the direction of propagation of the shear wave. They therefore have further to travel, and the shear wave should propagate at $\cos 45°$ times the velocity of a direct stress wave. This is approximately so.

Consideration of a wave of direct stress is more difficult. Detailed analysis of direct stress wave velocity shows that it depends not only on E but on Poisson's ratio v in two dimensions and on G in three dimensions, implying that the waves are not purely of direct stress but also contain a certain amount of shear. On this basis a complex stress wave appears to propagate in two stages. In the first, direct stresses propagate, accompanied by such shear stresses as are inherent in the direct stress part of the system. In the second, the remaining shear stresses propagate more slowly to complete the stress system.

There does not seem to be a general solution available (Kolsky, 1963, pp. 13–14).

(b) Stress concentrations

At low strain rates local concentrations of stress can occur where there are sudden changes in geometry such as holes or sharp notches (Ford and Alexander, 1977, pp. 299–300). One of the major uses of linear elastic stress analysis has been to study the effects of such stress raisers in producing stress concentrations; there is a large amount of published work on these.

At high strain rates other types of stress concentration can occur because of the reflection and interaction of stress waves. There does not seem to be a general theory of these concentrations.

(c) Common three-dimensional concepts

Invariants and deviators
The load deflection patterns which exist in a material are independent of the system of co-ordinates used to define them. Principal stresses and strains, for example, can be derived regardless of the co-ordinates chosen. It can be shown, in a more general way, that there are three stress invariants and three strain invariants which remain constant regardless of the co-ordinate system used. Any combination of these invariants is also invariant so that the relationships can be expressed in various ways (Ford and Alexander, 1977, pp. 31–7).

The three stress invariants are generally called J_1, J_2 and J_3, and the three strain invariants are generally called I_1, I_2 and I_3. Note that, in achieving the wide scope of invariants, detailed information is lost, and it is not possible to derive a unique stress system or set of principal stresses from the invariants.

Often it is convenient to consider three-dimensional stress systems from a false zero given by the mean direct stress. This is known as the *hydrostatic stress* $\bar{\sigma} = \sigma_{ii}/3 = J_1/3$, and is analogous to the pressure in a fluid. Stresses measured from the hydrostatic stress are called *deviatoric stresses* or *stress deviators*. They can be denoted by σ_i^1.

The strain produced by the hydrostatic stress is a volumetric strain known as the *hydrostatic strain* $\bar{\varepsilon} = \varepsilon_{ii}/3 = I_1/3$. Strains measured from the hydrostatic strain are called *deviatoric strains* or *strain deviators*, and can be denoted by ε_i^1.

For an isotropic material the hydrostatic stress can be shown to produce a change in volume but no change in shape, so that it is called the *dilatational strain* or *volumetric strain*. The deviatoric stresses, on the other hand, produce changes in shape with no change in volume and they can be shown to be entirely shear stresses. The strains they produce are called *distortional strains*. It is thus possible to separate out two distinct types of behaviour which can be analysed separately.

There are only two deviatoric stress invariants because, by definition, the deviator from J_1 is zero. They are generally called J_2^1 and J_3^1. Similarly there are two deviatoric strain invariants I_2^1 and I_3^1.

The elastic constants
It can be shown that, for a linearly elastic isotropic homogeneous material, only two elastic constants are needed to define its stress–strain relationships in three dimensions (Freudenhal, 1966, pp. 265–7). One of these covers distortion and the other volume change or dilatation. The constant for distortion is the shear modulus G, sometimes known as Lamé's constant μ. The constant for dilatation is expressed in two ways, either as Lamé's constant λ or as the bulk modulus K. It is not possible to measure λ, but it can be shown that

$$\lambda = \frac{vE}{(1+v)(1-v)}$$

so that it can be derived from simple test results. It is a compact way of expressing the stress–strain relationship and it is widely used in theoretical analysis. The bulk modulus is given by

$$K = \frac{\text{hydrostatic stress}}{\text{dilatational strain}}$$

$$= \frac{\bar{\sigma}}{\bar{\varepsilon}}$$

$$= \frac{J_1}{I_1}$$

It can be measured experimentally, and consequently it too is widely used.

Linearly elastic composite materials are not isotropic, so more than two elastic constants are needed to define their stress–strain relationships. The number needed depends on the degree of anisotropy, and the detailed analysis can become lengthy.

The elastic limit
There are two theories in common use for relating the yield stress obtained from a simple static tensile test to the yield stress under a complex stress system. Both make use of the observed fact that many materials become substantially incompressible at static yield so that no subsequent changes in volume occur. Consequently hydrostatic stress can be ignored and only deviatoric stresses need to be considered.

Both use the differences amongst the direct principal stresses to characterize the deviatoric stresses, and both give reasonable results for ductile materials which deform plastically after yielding. Neither is suitable for brittle materials which fracture at, or close to yield. These are discussed in Chapter 4 (Ford and Alexander, 1977, pp. 402–10).

One theory is commonly attributed to Tresca, the other to von Mises. The von Mises theory is the more elegant and appears to represent observed behaviour more accurately, but the differences between the two results are not large and the Tresca theory if often easier to apply (Calladine, 1969, pp. 49–55).

When wave propagation can be neglected there seems to be no good reason why the Tresca or von Mises theories should not be used with appropriate dynamic values of σ_y.

1.3.2 The wave equations

The wave equations are a well-known group of partial differential equations which are applicable to all types of wave. They can be studied at varying levels of complexity and solved in a number of ways to suit the problem being analysed, and they are of central importance in the detailed study of wave propagation. Published work is extensive and growing continuously. Even in the limited field of linear elastic stress waves it is almost impossible to keep abreast of it (Scott, 1978).

No attempt is made here to deal with the solution of wave equations as it is beyond the mathematical level of this book, and still presents a number of unsolved problems. Fortunately, a qualitative appreciation of the factors involved is often sufficient for carrying out practical work and understanding published results.

(a) A stress wave in a rod

Elementary theory
The approach uses the relationship force = mass × acceleration instead of the impulse–momentum approach given in Section 1.2. If a straight rod travelling at velocity v hits a stationary, rigid, flat surface the end of the rod stops suddenly. Eulerian co-ordinates are used. The behaviour of a small length of the rod δx long and originally distance x from the impact end is considered. Conditions are shown in Fig. 1.8.

At time t the element has moved distance u towards the impact end and the net force acting on it is

$$- A\left[\sigma - \left(\sigma - \frac{\partial \sigma}{\partial x}\delta x \right) \right] = - A\frac{\partial \sigma}{\partial x}\delta x$$

where A is the cross-sectional area, $-\sigma$ is the compressive stress and $\partial \sigma/\partial x$ is the rate at which stress varies along the rod. The mass of the element is $A\rho\delta x$ and its acceleration is $-\partial^2 u/\partial t^2$, where the direction of u is taken as positive towards the impact end. Equating force to mass times acceleration gives

$$- A\frac{\partial \sigma}{\partial x}\delta x = - A\rho\delta x\frac{\partial^2 u}{\partial t^2}$$

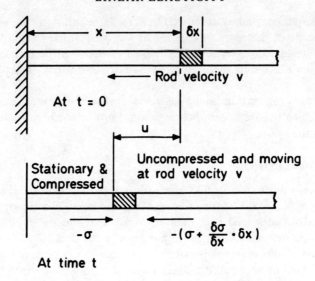

Figure 1.8 Linearly elastic rod striking rigid flat surface.

Therefore

$$\frac{\partial \sigma}{\partial x} = \rho \frac{\partial^2 u}{\partial t^2}$$

The compressive strain in the element is

$$-\varepsilon = \left(-\frac{\partial u}{\partial x} \delta x \right) \bigg/ \delta x = -\frac{\partial u}{\partial x}$$

Using the linear elastic relationship $\sigma = E\varepsilon$ gives

$$-\frac{\partial \sigma}{\partial x} = -E \frac{\partial \varepsilon}{\partial x}$$

$$= -E \frac{\partial^2 u}{\partial x^2}$$

Substituting for $\partial \sigma / \partial x$ gives

$$\rho \frac{\partial^2 u}{\partial t^2} = E \frac{\partial^2 u}{\partial x^2}$$

Hence

$$\frac{\partial^2 u}{\partial t^2} = C^2 \frac{\partial^2 u}{\partial x^2}$$

where $C = \pm \sqrt{(E/\rho)}$.

This is a simple wave equation; identical equations can be obtained for stress σ, strain ε and velocity v.

There are different types of general solution of the wave equations. The one which applies in this case is

$$u = f(x - Ct) + F(x + Ct)$$

where f and F are two arbitrary functions which need not be the same as each other. The solution represents two waves, both travelling along the rod at velocity C but in opposite directions. The direction can be found by plotting two values of x at two appropriate times on the characteristic plane. Taking $u = f(x - Ct)$ gives

$$u = f(x_1 - Ct_1) = f(x_2 - Ct_2)$$

and thus

$$C = \frac{x_2 - x_1}{t_2 - t_1}$$

Plotting x against t as in Fig. 1.9 shows that as t increases, x increases, so the wave is travelling in the positive direction and the solution applies along the characteristic line 0A. The wave in the negative direction, represented by $u = F(x + Ct)$, has no significance in the present case because only a stress wave in the positive direction is being considered.

Because f is an arbitrary function, the solution applies to an elastic stress pulse of any shape. The relationships $\sigma = \rho_0 Cv$ and $\varepsilon = v/C$ can also be derived.

Similar wave equations can be derived in two and three dimensions. The equations of motion are

$$\frac{\partial \sigma_{ij}}{\partial x_j} = \rho \frac{\partial^2 u_i}{\partial t^2}$$

Using the linearly elastic stress–strain relationships, the wave equation in

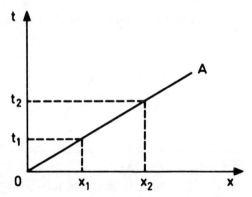

Figure 1.9 Direction of travel of simple stress wave.

three dimensions can be derived, and can be written neatly as

$$\rho \frac{\partial^2 u_i}{\partial t^2} = (\lambda + G)\frac{\partial I_1}{\partial x_1} + G\frac{\partial^2 u_i}{\partial x_{ij}}$$

where I_1 is the first strain invariant. In two dimensions a similar equation can be obtained, but it is more cumbersome to write because the condensed notation cannot be used fully. Neither equation appears to have been solved analytically except for the two limiting cases of direct stress with no shear and shear with no direct stress (Johnson, 1972, pp. 11–14). Shear stress propagates with a wave velocity of $C_T = \sqrt{(G/\rho)}$, as in the one-dimensional case. Direct stress in a two-dimensional sheet propagates with a wave velocity $C^1 = \sqrt{[E/\rho(1-\nu)^2]}$ and in a three-dimensional slab with a wave velocity $C^{11} = \sqrt{[(\lambda + 2G)/\rho]}$.

More exact theory
The elementary solution ignores transverse strains and the boundary conditions at the surface of the rod. Rigorous solutions for an elastic rod, impacted at one end, do not yet exist. The closest approximation is that due to Pocchammer and Chree, who analysed the behaviour of an infinite train of longitudinal sinusoidal waves in an infinitely long rod of circular cross-section (Kolsky, 1963, pp. 54–65).

The analysis is complex and a number of more approximate theories have

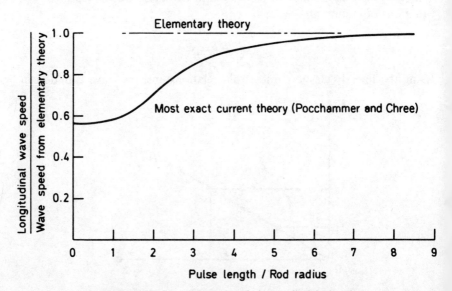

Figure 1.10 Stress wave speeds in a cylindrical rod.

been developed. All of the more exact solutions give a wave speed which varies with the ratio of wavelength to rod radius. As the stress pulse lengthens, the results tend to those of the elementary theory, which gives accurate wave speeds for pulses more than about ten times the rod radius. Some results are plotted in Fig. 1.10 (Davies, 1956).

The great complexity of the more exact solutions is the reason why condensed notation and condensed analysis are used so widely. Without them analysis would be intractable.

(b) Three-dimensional behaviour

There is a large and continuously increasing body of theoretical work on linearly elastic stress waves propagating in two and three dimensions. Most of this is not directly concerned with impact and all of it requires considerable expertise in mathematics or computing. A non-expert wishing to use this information is unlikely to get anywhere without expert assistance.

2

Non-linear and time-dependent elasticity

2.1 INTRODUCTION

2.1.1 General

(a) Types of behaviour

Brittle solids fracture at, or close to, the elastic limit with little or no inelastic deformation. Ductile solids continue to deform but inelastically and there is often a transition region in which both inelastic and elastic behaviour are important.

With rubber-like solids behaviour is substantially elastic up to large strains, where small deflection theory is no longer applicable but the stress–strain relationship is markedly non-linear. Elastic fluids also behave non-linearly at large strains. In addition there are strain rate, thermodynamic and temperature effects.

Departures from linear elasticity complicate analysis considerably and, because there are many different variations, the results from any one analysis are likely to be of limited application. Consequently, comprehensive general analysis of the type found with linear elasticity is not attempted and published work tends to be restricted to one dimension or a relatively simple set of conditions. A major feature of much of the analysis is the use of moduli which are analogous to Young's modulus but incorporate non-linear or time-dependent effects of various kinds.

(b) Transition zones

The properties of materials vary with both strain rate and temperature. With polymers there is such a consistent link between the two that they are often used interchangeably but, with other materials, the relationship is less clear-cut.

There are a number of temperature zones in which different types of behaviour occur. Some materials pass through all the stages but some miss one or more of them and there are large variations in the details of behaviour within the overall pattern. The effect of increasing strain rate is broadly the same as that of reducing temperature.

At low temperatures, materials are solid and brittle. As temperature increases there is a transition region leading to a zone of less brittle behaviour in which ductile or rubber-like behaviour occurs. As temperature rises further there is a second transition leading to a zone where the material is a liquid. This is followed by a third transition zone as the material changes to its final form as a gas.

A moderate change in strain rate in a temperature transition zone can alter the type of behaviour of the material, whereas a much larger change away from a transition zone can have a less marked effect. Well-known examples of these effects are the brittle behaviour of rubber or mild steel at low temperatures, and bouncing putty (silicone gum rubber), which bounces like a rubber ball on impact but flows like a liquid if left to stand for some time.

2.1.2 Basic relationships

(a) Creep and relaxation

Sometimes the behaviour of a material is time dependent. This can occur in various ways but it is usual to consider two limiting conditions – creep and relaxation. With *creep*, the stress is maintained constant but the strain continues to increase with the passage of time. With *relaxation*, the strain is held constant but the stress decreases with the passage of time, asymptotically approaching a limiting value.

(b) Damping

Not all of the work done by a stress system in deforming an elastic material is recoverable when the stress system is removed. There are always some losses and, though often these are small enough to be ignored, they sometimes have an appreciable effect. Usually the losses are said to be due to *internal friction* or *damping*.

Damping can be measured in a number of ways but for solid materials vibration tests are widely used because usually they are the easiest to carry out and to analyse. The values of damping obtained for one type of loading can be applied directly to other types of loading. The rate at which energy is dissipated by damping is generally taken to be proportional to the strain rate (Kolsky, 1963, Chapters 5 and 6).

Strain in a damped material occurs not at the same time as the stress which causes it but at a different time, which varies in a complex manner with the conditions of loading and the material properties. Instead of the simple Young's modulus E linking stress and strain it is necessary to use a complex modulus which has two components, one defining the elastic part of the behaviour and one the damping part.

2.2 MATERIAL PROPERTIES

2.2.1 Metals and crystalline solids

Some materials have a well-defined yield stress at which linearly elastic behaviour ends, but with others there is no clear-cut change. As stress increases there is a gradual increase in the proportion of the strain which is inelastic. A common method of dealing with this is to specify a notional yield stress; there are several ways of doing this, each of which is self-consistent, but the different versions can give markedly different values for the yield stress (Thwaite, 1968).

Time-dependent behaviour in the elastic–plastic region is negligible for metals under static loading except at elevated temperatures when long term creep occurs. In this case the loading times are hundreds to hundreds of thousands of hours and the associated strain rates range from about 10^{-9} to $10^{-12}\,\mathrm{s}^{-1}$ (McClintock and Argon, 1966, pp. 639–40).

Yield stress varies with strain rate in the way discussed in Chapter 1. When the material is linearly elastic, the very small changes which occur in Young's modulus with temperature and strain rate can usually be ignored. The effect of the increased yield stress is just to extend the upper limit of the linear stress–strain relationship (Fig. 2.1(a)). If the material behaves non-linearly then, as well as increasing the yield stress, the higher strain rate raises the stress–strain curve as a whole (Fig. 2.1(b)) in order to maintain a continuous curve.

In either case the change is of short duration. If the stress is maintained, the strain rises eventually to its static value and, if the strain is maintained, the stress falls eventually to its static value. As the duration of the load increases the appropriate stress–strain curve varies with the stage reached, moving lower with the passage of time until the static curve is reached. This behaviour can be regarded as visco-elastic.

In some metals, notably mild steel, an upper and lower yield stress exist under static loading, the stress dropping suddenly at constant strain when yield occurs. A similar drop in stress at yield is found in some metals of the same crystal structure at strain rates over $10^{3}\,\mathrm{s}^{-1}$, although it does not exist under static loading (Campbell, 1970, p. 71). In metals showing this drop there is a delay between the application of stress and the onset of yield. During this time the metal behaves elastically. The phenomenon has been studied extensively for mild steel and it has been found that the delay time varies with both stress and temperature. This is shown in Fig. 2.2 (Cottrell, 1957, Fig. 1.9). Under appropriate conditions the total delay time can be built up by successive applications of load each too brief to cause yield (Vigness et al., 1957, Fig. 4.5). A mild steel specimen might thus withstand an impact elastically several times and then yield without warning under a further, identical impact. This cumulative build-up disappears after a period of time. The phenomenon

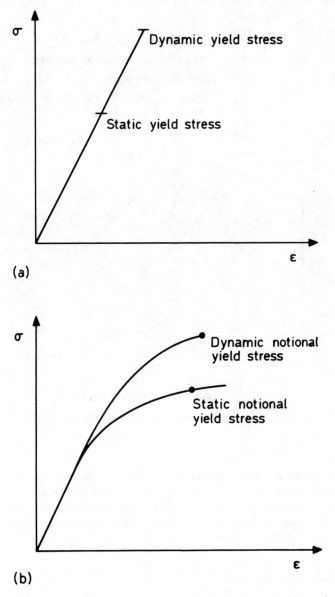

Figure 2.1 Enhanced yield stresses under dynamic loading. (a) Linear behaviour.
(b) Non-linear behaviour.

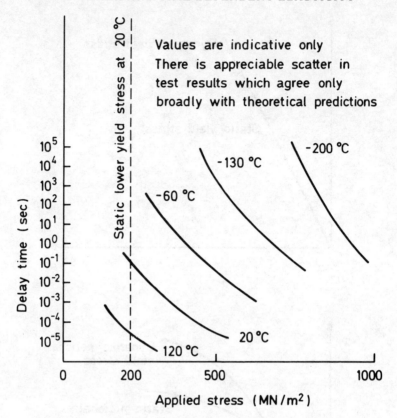

Figure 2.2 Variation in delay time to yield for mild steel.

is known as *ageing*, and the ageing time becomes shorter as temperature increases. At room temperature the ageing time is a few days.

Where a drop in stress occurs at yield, permanent strain can be distributed irregularly throughout the specimen. Crystal structure, ageing and irregular distribution of strain are discussed further in Chapter 3.

2.2.2 Rubbers and polymers

Rubbers and polymers have an internal structure which is completely different to that of metals and crystalline solids. They deform in a different way. The elastic strains can be large, the stress–strain relationship is markedly non-linear and behaviour in tension is markedly different from that in compression. As with metals, approximate moduli are used to define behaviour and these vary with the operating conditions which are expected.

Behaviour is basically elastic but an appreciable time-dependent element is also present. The material continues to deform slowly under a constant load and there is a delay in elastic recovery when load is removed. This behaviour varies greatly in detail and can be modified fairly readily to give desired properties.

The volume remains substantially constant during deformation so that rubbers and polymers can be regarded as incompressible compared with the strains occurring. Thus usually dilatational strains can be neglected and only deviatoric strains need be considered.

Materials which behave in a rubber-like way do so only over a limited range of temperatures and strain rates, as shown in Fig. 2.3. Because of the complex stress–strain behaviour of polymers it is not convenient to specify behaviour in terms of a representative stress, and the *relaxation modulus* E_R is frequently used instead. This is broadly analogous to Young's modulus E but varies with the nominal strain rate (McClintock and Argon, 1966, pp. 230–3). For a given material the strain rate and temperature effects on E_R are interlinked and can be used interchangeably. In fact graphs like Fig. 2.3 are constructed from data obtained over a limited range of strain rates and temperatures and then extrapolated to cover a wide range. It seems likely that the more extreme extrapolated values will be inaccurate because other factors are also likely to change so far away from the basic data.

Within the rubber-like region the elastic moduli change in a complex manner with both strain and strain rate. Dynamic moduli are higher than static ones but the relationships are not simple and need to be handled by damping or visco-elastic theories.

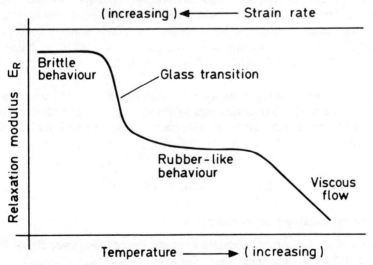

Figure 2.3 Temperature and strain rate equivalence for a polymer.

2.2.3 Fluids

Fluids can be either liquids or gases. Generally the bulk modulus of liquids is high and they can be regarded as approximately constant in volume except at very high pressures. When high pressures are applied the stress–strain relationship is non-linear in much the same way as the simple compressive relationship for a rubber.

The bulk modulus of gases is very low and they are readily compressed by large amounts. External work needs to be done to increase the pressure so that external energy is supplied to the gas. The additional energy increases the temperature of the gas and alters the relationship between pressure and volume.

There are two limiting conditions of behaviour. If the change in pressure occurs slowly the additional heat escapes as quickly as it is supplied and the temperature of the gas remains constant. This behaviour is known as *isothermal*. Volume always remains inversely proportional to pressure and, because all of the externally supplied energy is dissipated, it need not be taken into account in considering the behaviour of the gas. If the change in pressure occurs very rapidly, none of the additional heat has time to escape and all of the externally supplied energy remains in the gas. This behaviour is known as *adiabatic*. The temperature rises and the pressure–volume relationship is a power law of the general form

$$PV^\gamma = \text{constant}$$

where γ is a constant which is a characteristic property of the gas.

At low strain rates behaviour tends to isothermal, and at high strain rates to adiabatic. Attempts have been made to find analogous equations of state for liquids and solids. These have had only limited success, but the concepts of isothermal and adiabatic deformation are commonly used in analysing the behaviour of liquids and solids as well as gases (McClintock and Argon, 1966, pp. 190–4).

Liquids do not show the pronounced thermodynamic effects found in gases and their bulk modulus does not seem to change appreciably with strain rate, so probably strain rate effects on the elastic behaviour of liquids can be neglected.

2.3 VISCO-ELASTICITY

2.3.1 Linear behaviour in one dimension

In a linearly visco-elastic material linear elasticity and linear viscosity are both present. Behaviour of a linear visco-elastic material for direct loading in one dimension is considered here, but similar analyses apply to all stress systems.

For linear elasticity $\sigma/\varepsilon = E$ and for linear viscosity $\sigma/\dot{\varepsilon} = F$, where $\dot{\varepsilon}$ is the rate of strain and F is a viscous constant having the units of stress × time.

If an impact load is applied to the elastic component, stress and strain both jump instantaneously to their appropriate values. This implies an infinite strain rate on impact but, because strain rate is not specified in the relationship, no analytical problems arise.

If an impact load is applied to the viscous component, strain rate jumps instantaneously to the appropriate value, which is finite. Strain cannot jump to its appropriate value but has to build up from zero at the imposed strain rate so that, at the moment of impact, there is an inconsistency. This is overcome by applying a notional starting impulse known as a *Dirac spike*, consisting of an infinitely high stress acting for an infinitesimally short time to bring the strain to up to the appropriate value. The linear viscous relationship gives

$$\frac{d\varepsilon}{dt} = \frac{\sigma}{F}$$

By definition the notional applied impulse is $\sigma \, dt$ which is indeterminate, but this equals $F \, d\varepsilon$ which is known.

Elastic and viscous elements can be combined in various ways, a simple one being a single element of each type in series. Figure 2.4(a) shows two elements in series represented by a spring for the elastic one and a dashpot for the viscous one (McClintock and Argon, 1966, p. 244). This is known as a *Maxwell model*, and it behaves as a liquid in that inelastic strain continues to increase as long as stress is maintained.

The stress in both components is the same and the total strain is

$$\varepsilon = \varepsilon_S + \varepsilon_D$$

where ε_S is the strain in the spring and ε_D the strain in the dashpot. Differentiating with respect to time gives

$$\dot{\varepsilon} = \dot{\varepsilon}_S + \dot{\varepsilon}_D$$

$$= \frac{\dot{\sigma}}{E} + \frac{\sigma}{F}$$

This can be rewritten in various ways, but a common one is

$$\dot{\sigma} + \frac{1}{\lambda} = E\dot{\varepsilon} \tag{2.1}$$

where $\lambda = F/E$ and has the units of time. It can be regarded as a measure of the time the material takes to respond to an applied stress.

Although such models are useful for visualizing linear visco-elastic behaviour they do not represent adequately the behaviour of real materials, where relationships are more complex and λ varies with strain rate and

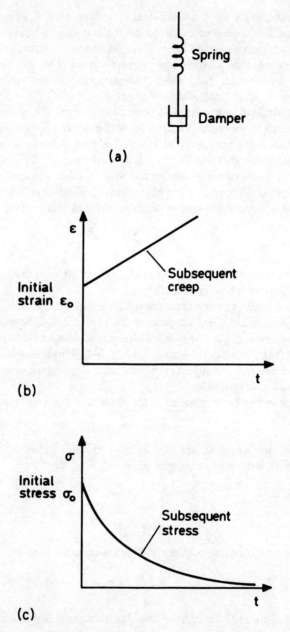

Figure 2.4 Maxwell visco-elastic model. (a) System. (b) Creep at constant stress. (c) Stress relaxation at constant strain.

temperature. Empirical values of creep compliance (the inverse of a creep modulus $E_C = \sigma/\varepsilon$) and relaxation modulus against strain rate are often obtained for rubbers and for metals at high temperatures. These can be measured directly if the loading times are sufficiently long but, for the rates of strain occurring in impact, they are usually derived from vibration tests (Kolsky, 1963, pp. 104–6).

Slightly more complex models incorporating up to two springs and two dashpots give better approximations over limited ranges of strain rates without being too complex to manipulate and understand. The values used for elastic and viscous behaviour in these models depend on the model chosen, so they cannot be regarded as intrinsic material properties analogous to Young's modulus.

2.3.2 More detailed analysis

Analysis of behaviour in three dimensions can become very complex, especially if dynamic effects and wave propagation are considered. The same equations define the stresses, strains, forces and accelerations in a material regardless of the stress–strain relationships which can be added separately. Techniques are in use for dealing with material properties in static analysis in three dimensions which can, in principle, be extended to impact analysis.

Hydrostatic and deviatoric stresses can be treated separately. In general, for a linearly visco-elastic material, the deviatoric behaviour can be taken to comply with the appropriate visco-elastic model. The hydrostatic behaviour can be dealt with in different ways, the most common being:

(1) The material is incompressible. This is the simplest model but it appears not to be generally applicable to visco-elastic solids.
(2) The material is linearly elastic. This has been widely used. Two interesting features arise from its use: all materials show an initial elastic response, regardless of the visco-elastic model chosen for the deviatoric stresses; and the ratio of transverse strain to direct strain varies with time so that Poisson's ratio is variable.
(3) The visco-elastic behaviour is the same in dilatation as in distortion. This is mathematically convenient and gives a constant Poisson's ratio. It becomes inaccurate when deformations are large because, in real materials, dilatations do not become nearly so large as deviations.

2.4 STRESS WAVES

2.4.1 Ignoring time-dependent effects

(a) Wave speeds
If a single stress–strain curve can be used throughout the analysis then simple linear theory can be extended to non-linear behaviour, and it can be shown

that the wave speed is given by (Kolsky, 1963, pp. 164–70)

$$C = \sqrt{\frac{E_{\text{tan}}}{\rho_0}}$$

where E_{tan} is the slope of the tangent to the stress–strain curve and is known as the *tangent modulus*. If E_{tan} decreases with increasing stress, then a stress wave disperses as it travels along, with the higher stresses travelling less quickly than the lower ones. If E_{tan} increases with increasing stress, higher stresses try to travel faster than lower ones. This is not possible and the wave sharpens up to a shock wave with the highest stress occurring instantaneously at the wave front.

(b) Dispersing waves

Figure 2.5(a) shows a stress–strain curve in which E_{tan} decreases with increasing stress. Figure 2.5(b) shows the variation in wave speed with stress derived from $C = \sqrt{(E_{\text{tan}}/\rho)}$. Figure 2.5(c) shows a fan of characteristic lines each representing the propagation of a single value of stress at its appropriate

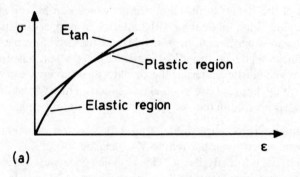

(a)

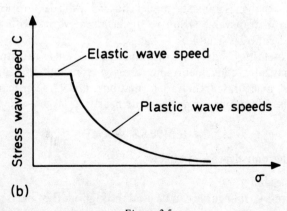

(b)

Figure 2.5

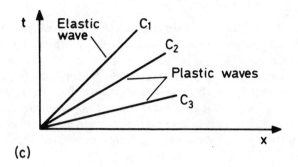

(c)

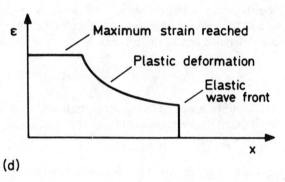

(d)

Figure 2.5 (a) Stress–strain curve in which E_{tan} decreases with increasing stress. (b) Variation in stress wave speed with stress. (c) Fan of characteristic lines representing different stresses. (d) Distribution of strain at time t.

wave speed. Because strain is assumed to occur instantaneously it travels at the same speed as the stress causing it. Figure 2.5(d) shows the distribution of strain at time t after the initial impact. The distinctive features of this analysis are the initial jump on an elastic wave front travelling at the initial elastic wave velocity, the constant amplitude of this elastic wave, and the plateau of constant strain spreading from the impact end once the maximum strain has been reached.

If the stress at the impact end is removed, an unloading wave travels along the rod at the elastic wave speed, overtaking (in time) all parts of stress wave except the initial elastic wave front, which is travelling at the same speed as the unloading wave (Kolsky, 1963, pp. 170–3). Plotting characteristic lines, it is straightforward to derive the residual permanent strain if elastic and inelastic strains can be separated (Zukas *et al.*, 1982, pp. 96–110).

(c) Shock waves

Figure 2.6 shows a stress–strain curve in which E_{tan} increases with increasing stress. In this case larger stresses try to propagate more quickly than smaller ones. This is not possible, and the wave sharpens up to a *shock wave* with the highest stress occurring at the wave front. To analyse this simply it is convenient to regard the material as behaving like a compressible gas. Generally shock wave equations are known as *Hugoniot equations* or *Hugoniots*. The derivation is analogous to that given for a wave of direct stress in a rod in Section 1.2.2, but it is slightly more complicated (Goldsmith, 1960, pp. 147–54).

Assuming isothermal behaviour,

$$P_0 V_0 = P_1 V_1$$

where P is pressure, V is volume, the subscript 0 indicates the region ahead of the wave front and the subscript 1 indicates the region behind the wave front.

The pressure difference across the wave front is

$$P_1 - P_0 = P_1 \frac{V_0 - V_1}{V_0}$$

If the cross-section of the material remains constant at A then, at time t after the start of the impact, the original volume of the compressed material is

$$V_0 = A C_s t$$

where C_s is the shock wave velocity. The compressed material behind the wave front has shortened by $(V_0 - V_1)/A = vt$, where v is the particle velocity.

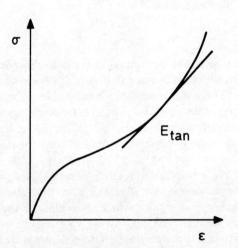

Figure 2.6 Stress–strain curve in which E_{tan} increases with increasing stress.

Eliminating A gives

$$\frac{V_0 - V_1}{V_0} C_s t = vt$$

or

$$C_s = \frac{V_0 v}{V_0 - V_1}$$

$$= \frac{P_1 v}{P_1 - P_0} \cdot$$

Equating impulse to momentum change,

$$(P_1 - P_0) A = \rho_0 C_s v A \tag{2.2}$$

where ρ_0 is the density of the uncompressed material. Hence

$$\frac{P_1 v}{C_s} = \rho_0 C_s v$$

giving

$$C_s^2 = \frac{P_1}{\rho_0}$$

Thus the shock wave velocity is

$$C_s = \pm \sqrt{\frac{P_1}{\rho_0}}$$

The particle velocity is

$$v = \frac{P_1 - P_0}{\rho_0 C_s}$$

$$= \frac{P_1 - P_0}{\sqrt{(\rho_0 P_1)}}$$

from which it follows that, as P_1 becomes much greater than P_0,

$$v = \frac{P_1}{\sqrt{(\rho_0 P_1)}} = C_s \tag{2.3}$$

In fact under shock wave conditions the material can be expected to behave adiabatically, so that it is necessary, in addition, to look at the energy balance. It is more convenient to work with densities; thus

$$\frac{v}{C_s} = \frac{V_0 - V_1}{V_0}$$

becomes

$$\frac{v}{C_s} = \frac{(1/\rho_0) - (1/\rho_1)}{1/\rho_0} = \frac{\rho_1 - \rho_0}{\rho_1}$$

The external energy supplied to the gas is given by

$$\text{external energy} = \text{external force} \times \text{distance it moves}$$
$$= P_1 A\, vt$$

Some of this is stored as kinetic energy in the moving column of gas and some goes to increasing the internal energy as the temperature rises. The mass of gas behind the wave front is $M = \rho_0 A C_s t$, so that the energy stored is

$$\rho_0 A C_s t \left(E + \frac{v^2}{2} \right)$$

where $1/2\, Mv^2$ is the kinetic energy and E is the increase in internal energy per unit mass.

Equating energy supplied to energy stored,

$$P_1 A vt = \rho_0 A C_s t \left(E + \frac{v^2}{2} \right)$$

Therefore

$$P_1 v = \rho_0 C_s \left(E + \frac{v^2}{2} \right)$$

$$E = \frac{P_1 v}{\rho_0 C_s} - \frac{v^2}{2}$$

Now

$$C_s = \frac{\rho_1 v}{(\rho_1 - \rho_0)} \qquad \text{and} \qquad v = \frac{P_1 - P_0}{\rho_0 C_s}$$

Therefore

$$E = \frac{P_1(\rho_1 - \rho_0)}{\rho_0 \rho_1} - \frac{(P_1 - P_0)(\rho_1 - \rho_0)}{2\rho_0 \rho_1}$$

$$= \frac{1}{2\rho_0 \rho_1} \left[(P_1 + P_0)(\rho_1 - \rho_0) \right]$$

$$= \frac{1}{2} \left[(P_1 + P_0) \left(\frac{1}{\rho_0} - \frac{1}{\rho_1} \right) \right] \tag{2.4}$$

This gives a total of three equations for the shock wave, relating particle velocity v, shock wave velocity C_s and increase in internal energy E to pressure and density. It is necessary to know seven parameters, $P_0, P_1, \rho_0, \rho_1, v, C_s$ and E in order to specify conditions completely. Usually P_0 and ρ_0 are known, and if any other two conditions are known the Hugoniots can be solved for the three remaining unknowns. The most commonly specified pairs of parameters are v and C_s or $(P_1 - P_0)$ and ρ_1, which can be measured fairly easily (Kinslow, 1970, pp. 295–6).

With many solids an approximately linear relationship exists between C_s

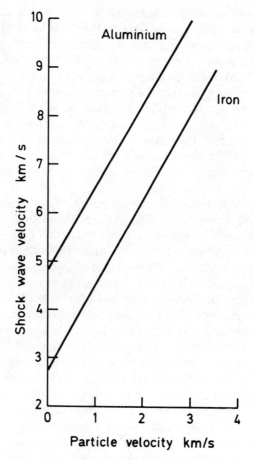

Figure 2.7 Hugoniot relationships for metals at very high strain rates.

and v of the general form

$$C_s = C + Sv$$

where C is the elastic wave velocity for a compressive stress wave and S is a constant which is a property of the material.

Typical Hugoniot relationships are shown in Fig. 2.7 (Kinslow, 1970, p. 317).

2.4.2 Time-dependent effects

(a) Linear visco-elastic waves

If the stress–strain relationship for a material changes during the propagation of a stress wave, this modifies the behaviour of the wave. Simple linear visco-elastic models can be studied by computer and, although analysis can become very complicated, it is fairly easy to visualize the results qualitatively. There is not a constant relationship between stress and particle velocity (Kolsky, 1963, pp. 116–22), and it is convenient to consider two limiting conditions for the impact of a moving rod with a rigid stationary anvil. In the first of these, the relative velocity between the moving part of the rod and the anvil remains constant but the stress at the impact end of the rod is free to vary with time. This is analogous to relaxation and it appears to be a practical loading condition. In the second, the stress at the impact end of the rod remains constant, but the velocity of the moving part of the rod is free to vary with time. This is analogous to creep and it does not seem to be a likely loading condition in practice.

Four variations in material response are possible. Two depend on whether the material behaves basically as a liquid, with no limiting value of strain for a given stress, or basically as a solid with a finite strain for a given stress. The other two depend on whether the initial impact response is governed by a spring or a dashpot. There are thus two types of loading condition and four types of material response, giving eight basic variations in all, so that even a simple discussion can be confusing (Lee and Morrison, 1956).

For both constant velocity and constant stress inputs to a visco-elastic solid, the stress behind the wave front builds up to a finite value and remains there as long as the input is maintained. This is generally similar to the behaviour of a simple solid whose properties do not vary with time, though there are differences in detail. For a constant velocity input to a visco-elastic liquid the stress behind the wave front builds up to a maximum value and then decays exponentially to zero, and the wave itself decays as it travels along. For a constant stress input to a visco-elastic liquid, stress falls with distance from the impact end, but the existence of a constant stress at the impact end prevents it from decaying to zero. To maintain this constant stress the moving part of the rod needs to move at a constantly increasing velocity relative to the anvil.

If the initial response is governed by a spring, initial stress propagates at the elastic wave speed. The amplitude of the elastic stress at the wave front decays exponentially with distance along the rod and, eventually, the initial elastic stress can be neglected. Once this has occurred stress can be regarded as building up smoothly from zero as the wave front passes, and the wave can be regarded as travelling at some appropriate velocity less than the elastic wave speed.

If the initial response is governed by a dashpot, a Dirac spike is needed to provide the appropriate initial conditions for analysis. Momentarily stress is infinite and strain is zero; the material consequently acts as though it were infinitely stiff, and the initial wave velocity is infinite. The rod is stressed instantaneously all along its length and stress everywhere starts to build up from zero. The rate of build-up falls off rapidly with distance from the impact end, however; the resulting behaviour can be regarded as very similar to that of an initially elastic material.

As with other linear visco-elastic analyses, these models of stress wave propagation are useful for visualizing behaviour but do not represent adequately the behaviour of real materials. Three- or four-parameter models of the material can give better results but the analysis and interpretation get more difficult.

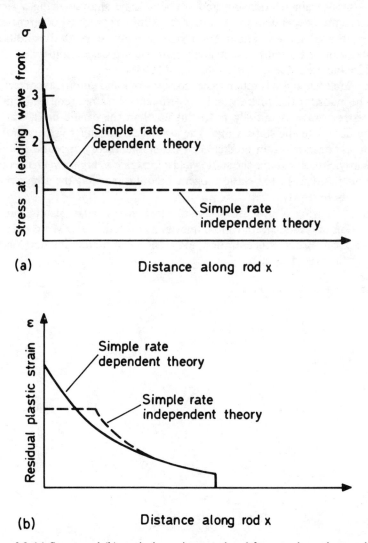

Figure 2.8 (a) Stress and (b) strain in an impacted rod for rate-dependent and rate-independent theories.

(b) Waves in real materials

Simple stress wave theory, ignoring time-dependent changes in material behaviour, is unlikely to give useful results for rubbers, where such effects are marked. On the other hand, although there are time-dependent effects in metals stressed above the static elastic limit, these are often not large enough to affect behaviour greatly. Figure 2.8 shows theoretical results for the constant velocity impact of an aluminium rod using both a simple analysis based on the static stress–strain relationship and a visco-elastic analysis using a simple model of a time-dependent strain rate effect. Although there are differences the two results are basically similar, and it may well not be worth the additional complication of the more exact analysis in such a case (Goldsmith, 1960, Figs 125 and 127; Zukas et al., 1982, pp. 110–31).

The simpler analysis is much easier to carry out and empirical corrections could be made for the differences. In experiments the amplitude of the initial elastic stress wave is initially much higher than the elastic limit, but this quickly decays to the static value. The strains do not differ greatly and the plateau of constant strain predicted by the simple theory appears, although less clearly. Stresses above the static yield propagate at higher velocities than those predicted from the static tangent modulus, so that the stress pulses disperse more slowly.

Metals showing a drop in stress at yield give similar results but the distribution of plastic strain can be uneven as in static tests. More than one area of local strain concentration may be found under impact loading (Goldsmith, 1960, Fig. 207).

3
Plasticity

3.1 INTRODUCTION

3.1.1 General

(a) Plastic flow

The inelastic deformation of ductile solids can vary with stress distribution, previous loading history and temperature as well as with strain and strain rate. In addition, for a given overall load and deformation, strain can be distributed either uniformly or nonuniformly. Thus two lines of approach are needed; one deals with average strains, and the other with local concentrations of strain.

The inelastic deformation process is called *plastic flow*. At low strain rates it consists almost entirely of shear deformation and the process is broadly similar to the flow of an incompressible fluid of high viscosity. At higher strain rates, inertia forces become progressively larger relative to the shear strength of the material, until at very high strain rates they are so large that shear strength is negligible but compressibility is important. The flow process is then similar to the flow of a compressible fluid of zero viscosity.

The stress at which flow occurs is called the *flow stress*. In some materials the flow stress, at the lower strain rates, remains substantially constant, regardless of strain, but in others strain hardening occurs, the flow stress increasing with increasing strain. Strain softening can also occur.

Commonly used models of ideally plastic materials incorporate five basic assumptions:

(1) Yield occurs sharply, and so the onset of plastic flow is clearly defined.
(2) Tension and compression are identical apart from a change of sign.
(3) The material is isotropic with identical properties in all directions.
(4) The stress–strain curve can be approximated by a few straight lines.
(5) The material is incompressible.

The last of these considerably simplifies analysis in three dimensions and, except at very high strain rates or pressures, the compressibility of solids is so small that it can be neglected. If follows that plastic strain can usually be regarded as being entirely deviatoric and consequently dilatational strain can be ignored (Ford and Alexander, 1977, pp. 394–400).

Curves of static tensile strain are derived from the overall load and overall extension. They imply that plastic deformation is distributed uniformly throughout the specimen. This is not necessarily so, and local concentrations of plastic strain can occur in two different ways even in a simple static specimen.

One of these occurs only in tension but is common to all ductile materials. As the specimen stretches the cross-sectional area reduces and there comes a stage at which the cross-sectional area is reducing faster than the material is strain hardening. If there is no strain hardening this occurs as soon as the material starts to flow plastically. The load needed to keep the specimen stretching then falls steadily with increasing plastic strain. The stress at which the overall tensile load on the specimen is at a maximum is called the *ultimate tensile strength* of the material.

Because of local imperfections the reduction in load occurs first at a single localized position. As the load falls there, the rest of the specimen is unloaded and so does not again reach the ultimate tensile strength. All further plastic strain is concentrated at the initial location and the cross-section there reduces to a narrow neck which ultimately fractures (McClintock and Argon, 1966, pp. 320–5).

The other type of local concentration of plastic strain occurs in all types of loading but only in materials which strain soften. Again, because of local imperfections, strain softening occurs first at a single location and further plastic flow is concentrated there. A common case of this is the drop in stress at yield which occurs with mild steel. In this case the material starts to strain harden again, so the process could keep repeating at different locations until the whole specimen had passed through the strain softening stage. In practice plastic deformation spreads along the specimen from the initial location under an approximately constant load so that usually the upper yield stress can be detected only once at the initial yield (McClintock and Argon, 1966, pp. 311–16).

(b) Crystal lattices and dislocations

The atoms inside the crystals of a crystalline solid are arranged regularly in three-dimensional patterns known as *crystal lattices*. An element in a given state has only one type of lattice. There are three possible types, known as *face-centred cubic*, *body centred cubic* and *hexagonal close packed*, and these have such a large effect on behaviour that it is often more useful to compare crystalline materials by lattice type rather than, say, by whether they are metals or non-metals. When compounds or alloys are formed from a combination of different elements some distortion of the basic lattice patterns can occur, and this contributes to the changes in material properties which are found (McClintock and Argon, 1966, pp. 17–20).

In real materials small defects exist in the crystal lattices; these defects are known as *dislocations*. Plastic deformation of crystalline materials can be described in terms of dislocations and their behaviour. Dislocations can come in a number of different types each of which produces a distinctive elastic stress field round it. If the dislocation moves, parts of the lattice need to move to allow it to pass, and this produces an inertia effect which is the same as if the dislocation itself possessed mass. If the movement is caused by external forces applied to the crystal this can similarly be regarded as equivalent to a set of forces applied directly to the dislocation. It is convenient in analysis to regard the dislocation as if it were a small body with a known mass and subjected to applied forces (McClintock and Argon, 1966, p. 111).

3.1.2 Basic relationships

(a) Definitions of strain rate

Plastic deformation is a process of continuous flow and it is often more convenient to consider incremental changes of strain rather than total strain from some initial condition. It is common practice to consider a rate of change of strain in the general form $d\varepsilon/dx$, where x is some appropriate parameter. Usually under static loading this parameter is stress or distance, giving strain in terms of $d\varepsilon/d\sigma$ or $d\varepsilon/dl$. Often for simplicity the material is assumed to be flowing at a constant rate so that dl/dt is unspecified but constant. In these circumstances a shorthand notation is often used with $d\varepsilon/dl$ being written as

$$\frac{d\varepsilon}{dl}\frac{dl}{dt} = \frac{d\varepsilon}{dt} = \dot{\varepsilon}$$

This is identical to the notation for the rate of change of strain with time used in the study of impact. The term 'strain rate' or 'rate of strain' is also identical, so confusion can exist (Ford and Alexander, 1977, p. 480). It should be remembered that, with plastic flow, the expression 'high strain rate' can mean either a high velocity of deformation or a rapid change of strain with distance at an unspecified constant velocity, which is usually very low. Throughout this book the expression 'strain rate' and the notation $\dot{\varepsilon}$ mean only velocity of deformation $d\varepsilon/dt$ unless a specific exception is made.

Attempts have been made to relate strain rate effects in plastic flow to the dynamic behaviour of dislocations. A simple form of this is

$$\dot{\varepsilon} = \frac{\dot{\sigma}}{E} + \alpha\rho v$$

where ρ is the density of mobile dislocations (i.e. number per unit volume), v is the average dislocation velocity and α is a constant related to dislocation geometry. Such equations are difficult to apply because ρ and v are not easy to measure in dynamic loading.

Another, more general form uses the second invariants of the deviatoric stress J_2^1 and the deviatoric plastic flow I_2^{1P}. This is expressed as $I_2^{1P} = f(J_2^1)$ where f is some appropriate function. The field of dislocation dynamics is active but no clear-cut results seem to have emerged yet.

The movement of dislocations is assumed to be impeded by barriers which can be overcome by the supply of sufficient energy. This energy can come from either a stress or a thermal energy which is dependent on temperature. Theoretical results can be correlated with the observed behaviour of materials in tests. This correlation can involve extrapolating theoretical values of $\dot{\varepsilon}$ in the range $10^{10}\,\text{s}^{-1}$ to $10^{20}\,\text{s}^{-1}$, which are many orders of magnitude greater than the values as defined in this book. The relationships are quite complicated and beyond the scope of this book.

Various attempts have been made to obtain mechanical equations of state in which both strain rate and temperature effects can be interrelated. One approach, which has been used in the analysis of creep behaviour, uses a *velocity modified temperature*, which can be expressed in the general form

$$T_M = T\left(1 - \alpha \log\frac{\dot{\varepsilon}}{\dot{\varepsilon}_0}\right)$$

where T_M is the modified temperature, T is the temperature measured from absolute zero, $\dot{\varepsilon}$ is the strain rate and α and $\dot{\varepsilon}_0$ are material constants. The method appears to work only at low strain rates and temperatures below the recrystallization temperature (McClintock and Argon, 1966, pp. 190–4) see Section 3.2.3(a).

Another approach, which seems to be suitable at higher strain rates and temperatures, is to use the *Zener-Holloman parameter Z*. This is defined as

$$Z = \dot{\varepsilon}\,e^{Q/RT}$$

where Q and R are material constants. Z is in units of s^{-1}, so it can be regarded as a type of strain rate, but even at normal hot working strain rates and temperatures it can be in the range $10^{10}\,\text{s}^{-1}$ to $10^{20}\,\text{s}^{-1}$. This is many orders of magnitude greater than the values of $\dot{\varepsilon}$ as defined in this book.

(b) Interaction between strain and wave propagation

The time taken to reach a given strain varies inversely with the strain rate. At low strain rates the loading time is long compared with the propagation time of elastic waves, so wave effects can be neglected. At higher strain rates, say between $10^2\,\text{s}^{-1}$ and $10^4\,\text{s}^{-1}$, the effects of wave propagation can be separated out and the material properties can be stated without direct reference to them.

At very high strain rates, say $10^5\,\text{s}^{-1}$ and above, shock waves occur. It is no longer possible to separate material properties from stress wave propagation and results need to be presented in the form of Hugoniots.

In the analysis of plastic flow, materials are generally assumed to be incompressible. This leads to an infinite theoretical wave speed for dilatational waves. Simple models of ideally plastic materials often assume that the material is rigid up to yield, and this gives an infinite theoretical wave speed for elastic waves. Such models often also assume a constant flow stress, and this gives a zero theoretical wave speed for plastic stress waves.

If wave speeds need to be taken into account, it is necessary to make some more realistic assumptions about stress wave speeds. Common sense indicates that elastic waves should be regarded as propagating at the appropriate elastic wave speed. The selection of an appropriate wave speed for plastic waves is more difficult and depends on whether significant time-dependent effects are likely to be present. If they are not, the wave speed can usually be taken as the value given by using the tangent modulus; if they are, it will probably be necessary to use an appropriate visco-elastic wave speed. Unloading waves are generally assumed to travel at the elastic wave speed.

3.2 PLASTIC FLOW

3.2.1 Non-uniform strain

In static and low speed tests it is possible to strain a specimen at a constant rate. This becomes progressively more difficult at higher speeds as transient dynamic effects become more important, but at rates up to, say, $10^4 \, s^{-1}$ it is often possible to apply a constant load. In this case the strain rate varies throughout the test as in Fig. 3.1, but usually only a single strain rate is quoted

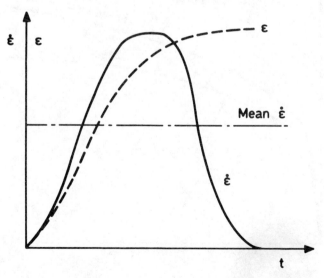

Figure 3.1 Strain rate occurring at constant stress.

which is an average value such as the total strain divided by the total time at load. This is acceptable because, in most cases, it is only the order of magnitude of the strain rate which is significant.

At still higher speeds, initial velocities and overall energy inputs can be controlled but both load and strain rate vary throughout the test.

The necking which is found in static tensile specimens also occurs in dynamic tensile tests. Sometimes more than one neck is present at higher strain rates, so it appears that the time to initiate necking is roughly comparable with the time for the stress in the specimen to fall below the dynamic ultimate tensile strength (Johnson, 1972, Plate 6). If the impact velocity is above a critical value, a tensile specimen breaks at once at the impact end. Simple stress wave theory, ignoring time effects, gives this critical velocity as the value at which the stress–strain curve becomes horizontal and the wave speed becomes zero. In tests the critical velocity is generally higher than this by a factor of between 1 and 3, which agrees with yield and ultimate stresses increasing with high strain rates by up to a factor of 3 (Kolsky, 1963, p. 185).

At strain rates over $10^3 \, s^{-1}$ the upper and lower yield stress phenomena can occur in metals with body centred cubic lattices where it is not found statically (Campbell, 1970, p. 71) and, presumably, this could lead to non-uniform distribution of plastic strain. Strain softening can also occur because of adiabatic heating. If a high local strain rate starts, heat is generated which has

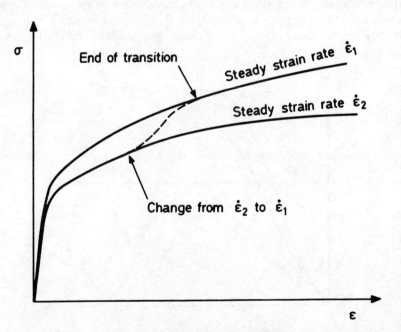

Figure 3.2 Effect of loading history.

no time to escape. This lowers the local flow stress so that very large strains occur locally whilst the rest of the specimen is substantially undeformed (Cottrell, 1957, Fig. 1.2). The effect of temperature is discussed further in Section 3.2.3(a).

If the strain rate is changed suddenly during a test the resulting flow stress is usually between the original value and the value which would have been found if the second strain rate had been applied throughout. The flow stress then changes with increasing strain until it reaches the value appropriate to the second strain rate. A typical example of this behaviour is shown in Fig. 3.2 (Campbell, 1970, pp. 52–6). Behaviour is not always consistent and sometimes the flow stress overshoots the value appropriate to the second strain rate, although it still comes back to this value as strain increases. There is some evidence to suggest that such overshoots may be spurious.

As with static loading there is an ageing process and the effect of loading history disappears after a time which depends on the temperature.

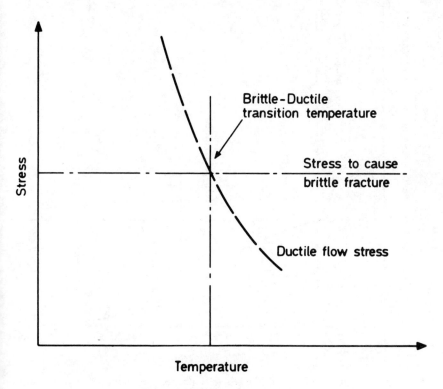

Figure 3.3 Davidenkov diagram.

3.2.2 Temperature and pressure

(a) Temperature effects

A metal fractures in a brittle manner or flows plastically depending on which mode of deformation requires the lower stress. Flow stress increases with decreasing temperature but the stress to cause brittle fracture seems to be much less dependent on temperature. Consequently, if the temperature is lowered there comes a stage at which brittle fracture occurs rather than plastic flow. Anything which increases the flow stress raises the temperature at which this change in behaviour occurs. This is illustrated in Fig. 3.3 (McClintock and Argon, 1966, Fig. 17.20) and is discussed further in Section 4.2.

At high temperature, plastic strains can be very large, and usually torsion tests are used to measure flow stress and ductility because many revolutions of twist can be obtained without greatly altering the overall dimensions of the specimen. This makes it difficult to estimate the flow stress, and often only comparative tests are carried out with the results given simply as torque and number of revolutions.

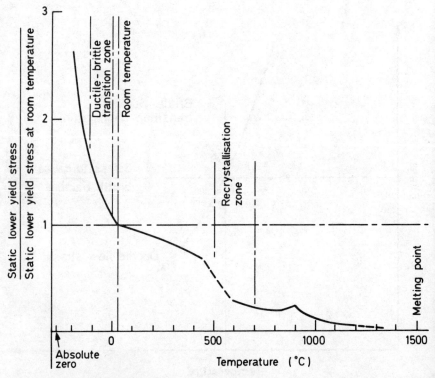

Figure 3.4 Variation in lower yield stress for mild steel with temperature under static loading.

At a high temperature – often roughly half the melting temperature measuring from absolute zero – the crystals in the metal reform after distortion. This is known as the *recrystallization temperature*. At the recrystallization temperature ageing takes place almost immediately although the mechanism by which this occurs is not necessarily directly linked to the recrystallization process (Stüwe, 1968).

Figure 3.4 gives a broad indication of how static yield stress varies with temperature, but there can be variations in detail for different materials and between the results obtained by different experimenters. Flow stress follows a similar pattern but it is harder to quantify results from different sources (McClintock and Argon, 1966, Fig. 17.3; Campbell, 1970, Fig. 13; Wagenaar, 1968).

(b) Compressibility at very high pressures

Materials which are usually regarded as incompressible are compressed by significant amounts at pressures of the order of $10^6 \, \text{kN m}^{-2}$. Such very high pressures occur statically in the deep sea and in subterranean rocks and can be produced statically in the laboratory. They also occur dynamically at strain rates of the order of $10^5 \, \text{s}^{-1}$ (Goldsmith, 1960, Fig. 77). Tension stresses of this magnitude do not occur.

3.2.4 Variation in flow stress with strain rate

(a) Room temperature

As strain rate increases, the ultimate tensile strength of a tensile specimen usually increases (Fig. 3.5(a)) (Campbell, 1970, Fig. 6; Goldsmith, 1960, Figs 245, 246, 254) and occurs at a lower strain (Fig. 3.5(b)) (Campbell, 1970, Fig. 6). Estimating the strain at which the ultimate tensile strength is reached is a matter of judgement: thus Fig. 3.5(b) is only approximate, but it implies that the ultimate tensile strength would be reached at zero strain when the strain rate is about $10^3 \, \text{s}^{-1}$.

The flow stress in a simple test specimen also increases with increasing strain rate. The mechanism of deformation appears to be broadly similar all the way from the very low strain rates found in creep up to rates of about $10^2 \, \text{s}^{-1}$ to $10^3 \, \text{s}^{-1}$. Throughout this range the flow stress increases approximately linearly with the logarithm of the strain rate. The shape of the stress–strain curve remains much the same but the increase in flow stress with strain rate tends to be less than the increase in yield stress, as is shown in Fig. 3.6 (Campbell, 1970, Figs 53, 56). Once plastic flow is established the increase in flow stress appears to be broadly independent of strain, as can be seen in Fig. 3.7 (Campbell, 1970, Fig. 56).

At strain rates over about $10^3 \, \text{s}^{-1}$ behaviour begins to change and flow stress increases more quickly, though there is some evidence that this change in

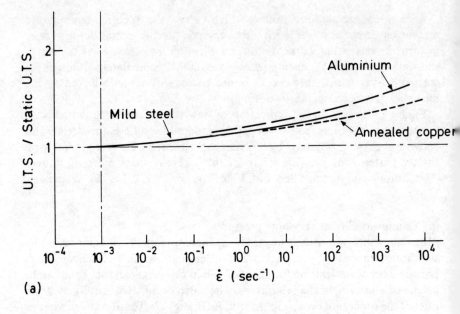

(a)

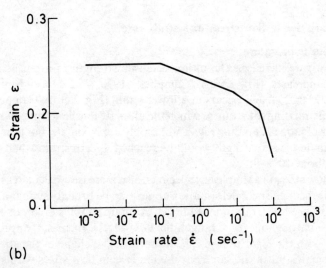

(b)

Figure 3.5 (a) Variation in ultimate tensile strength with strain rate at room temperature. (b) Strain at which UTS is reached.

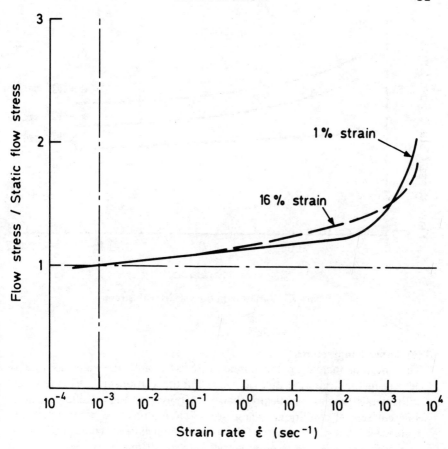

Figure 3.6 Variation in flow stress with $\dot{\varepsilon}$ at room temperature.

behaviour does not always occur or even that it may be a spurious result arising from the test method used. If the experimental curves are extrapolated the flow stress tends to infinity at rates of about $10^5 \, s^{-1}$ to $10^6 \, s^{-1}$, implying an upper limit to the rates of strain possible in solids. At such strain rates, however, the stresses are so high that the resistance of the material to distortion becomes negligible and its compressibility becomes important. In this state it is akin to a compressible fluid of zero viscosity, the concepts of stress and strain in a solid no longer apply, and hydrodynamic equations of state need to be used.

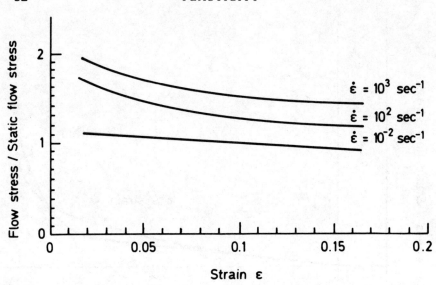

Figure 3.7 Variation in flow stress with strain.

(b) Other temperatures

The effect of increasing the temperature is broadly the same as that of decreasing the strain rate, and vice versa, but the relationship is less consistent with metals flowing plastically than with polymers and is different under different sets of conditions. These sets of conditions can be grouped into regions within which behaviour follows a common pattern. There appear to be four or five regions, but the boundaries between them are not clearly defined and large overlaps occur. In region I, which covers temperatures from about room temperature to about 500 °C and strain rates up to about $10^2 \, s^{-1}$, yield and flow are fairly insensitive to both temperature and strain rate. In region II, which covers temperatures from about -200 °C to about 100 °C and strain rates up to about $10^4 \, s^{-1}$, yield and flow are appreciably more sensitive to temperature and strain rate. Region III covers extremely low temperatures from absolute zero (-273 °C) to perhaps about -100 °C and does not seem to have been widely investigated experimentally. Region IV covers the rest of the possible sets of conditions including both very high strain rates and very high temperature, and behaviour in this region is very sensitive to temperature and strain rate. For some materials region IV is subdivided into a lower strain rate region IVa and a higher strain rate region IVb.

Figure 3.8(a) shows results for the lower yield stress of mild steel specimens loaded in shear (Campbell, 1970, Fig. 13). In this case results are in regions I, II and IV. In regions I and II the yield stress increases linearly with the logarithm of the strain rate but in region IV the rate of increase is much larger

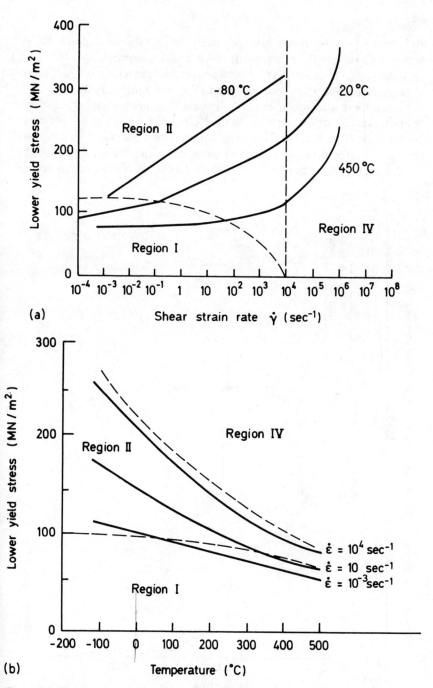

Figure 3.8 Temperature/strain-rate regions for mild steel in shear. (a) Strain rate plots. (b) Temperature plots.

and appears to be rising asymptotically to infinity at around $10^6 \, s^{-1}$. Figure 3.8(b) shows the same results plotted against temperature. In region I yield stress does not change much with temperature, in region II it falls rapidly as temperature increases, and in region IV it falls more rapidly still.

Figure 3.9 is a replot of Fig. 3.4 showing the approximate effect of both temperature and strain rate for mild steel (Campbell, 1970, Fig. 13; Wagenaar, 1968; Oxley, 1974, Fig. 14). There are two transition zones in which the rapid change in material properties with temperature leads to a correspondingly rapid change with strain rate. One is the ductile–brittle transition zone, in

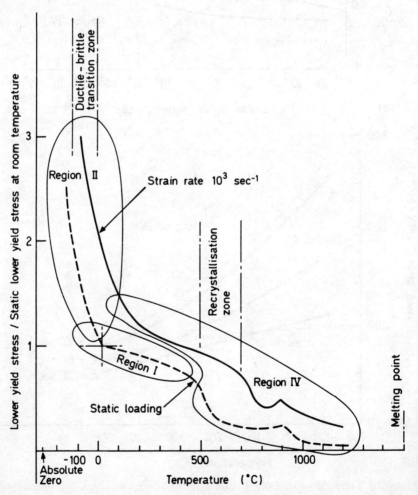

Figure 3.9 Variation in lower yield stress for mild steel with temperature at $10^3 \, s^{-1}$ strain rate.

which a modest increase in strain rate can raise the yield and flow stresses markedly. The other is a zone at about 900 °C in which the yield stress increases with increase in temperature. The latter effect can be significant in the hot working of metals (Wagenaar, 1968, Fig. A6).

Generally a metal becomes more ductile with increasing temperature, but a phenomenon called *hot shortness* can occur where the metal becomes relatively brittle at a temperature well above the recrystallization temperature (Harris and Barnard, 1968).

In adiabatic deformation, which is more likely at higher strain rates because the heat generated by internal friction has no time to escape, increasing strain rate can cause a considerable increase in temperature in a localized area. This can lead to failure with very little overall plastic flow but intense local plastic flows in small areas, leading to a clean fracture with very little other deformation and leaving changes in the structure of the metal which indicate the high temperature reached. Examples are high speed punching and the penetration of armour by shot.

If a rectangular block is compressed rapidly between parallel platens, the plastic work and the resulting heat are not spread uniformly throughout the block but tend to be concentrated along the arms of a cross which runs diagonally from corner to corner of the block. Appreciable increases in temperature can occur in these areas, which are known as *heat lines*. Heat lines of more complex shape can be found in other loading conditions, and these can cause adiabatic softening and metallurgical changes (Johnson, 1972, p. 208).

3.3 MORE COMPLEX ANALYSIS

3.3.1 Behaviour in two dimensions

There are two main approaches to studying plastic flow in more than one dimension. In the first of these, simplifying assumptions are made about the flow process. Usually a single average strain rate is assumed throughout the material; this is taken to be the overall strain divided by the total time it takes to occur. With such a simple model of flow it is possible to make allowances for strain hardening and the effects of strain rate and temperature.

Various formulae have been proposed in which strain hardening can be represented by a reasonably tractable mathematical relationship. One of the simplest is

$$\sigma = \sigma_1 \varepsilon^n$$

where σ and ε are one-dimensional stress and strain and σ_1 and n are material constants which can vary with strain rate and temperature (Campbell, 1970, Fig. 61).

In the second method simplifying assumptions are made about the material properties but the flow pattern is studied in detail. If conditions are restricted to plane strain and quasi-static loading then a *slip line field* analysis can be used (Ford and Alexander, 1977, Chapter 30). Slip line field analysis does not seem to be practicable, at present, if material properties vary with strain rate, but flow patterns obtained from the quasi-static slip line field analysis of a simplified material can be used as a basis for approximate analysis of more realistic behaviour.

Tests of specimens under combined loading appear to have been restricted to two axes, usually the combined tension and torsion of a hollow cylindrical specimen. In presenting these results it is necessary to reduce them to an effective stress σ_{eff} and an effective strain ε_{eff} using an appropriate analysis of the general type outlined in Chapter 1. Sometimes σ_{eff} and ε_{eff} are given as $NJ_2^{\frac{1}{2}}$ and $NI_2^{\frac{1}{2}}$, where $J_2^{\frac{1}{2}}$ and $I_2^{\frac{1}{2}}$ are the second invariants of deviatoric stress and strain. Typical results for aluminium are given in Figs 3.10 and 3.11 (Campbell, 1970, Figs 8 and 9).

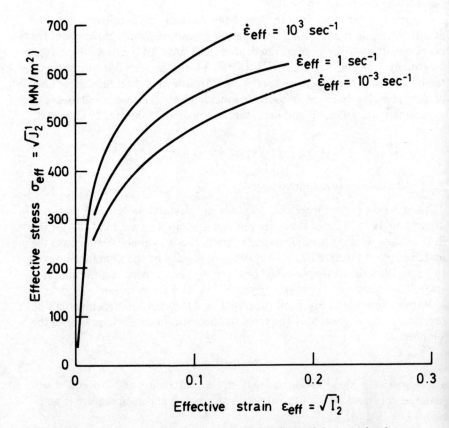

Figure 3.10 Combined loading behaviour of aluminium: strain plots.

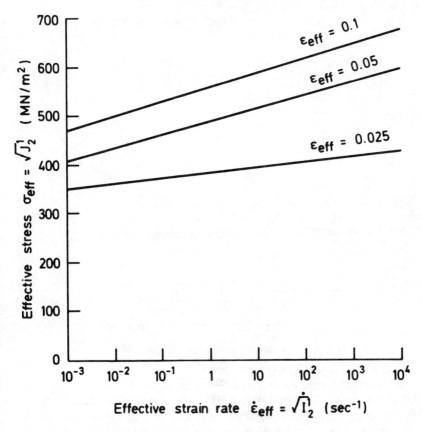

Figure 3.11 Combined loading behaviour of aluminium: strain rate plots.

Under static loading the magnitude and direction of the principal stresses can be varied during the test. Under dynamic loading the strain rates can be altered as well, so published results can be complex and need careful study.

At very high strain rates $(10^5\,\text{s}^{-1})$ the flows are very large and neither a simple Eulerian nor a simple Lagrangian system of co-ordinate axes can cope. This has led to the development of complex definitions of flow which are unsuitable for simple analysis but can be used successfully in a computer analysis (Zukas *et al.*, 1982, Chapter 10).

3.3.2 Microscopic effects

Changes in microstructure occur during plastic flow. These can be studied qualitatively and quantitatively after the impact and used as a basis for analysing the effects of strain rate and temperature.

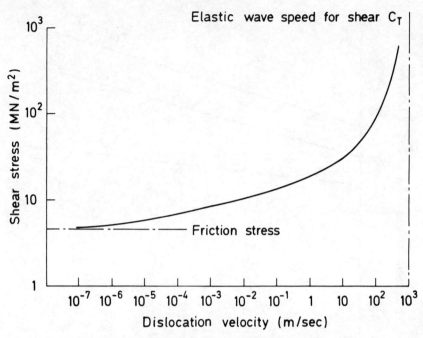

Figure 3.12 Dislocation velocities in lithium fluoride.

A minimum stress, known as the *friction stress*, is necessary to make a dislocation move. This can vary considerably with the type of dislocation and the type of crystal lattice. If the stress is increased the dislocation moves with a velocity which increases with the applied stress up to a limiting velocity at which the stress required tends to infinity. Figure 3.12 shows approximate dislocation velocity against applied stress for a lithium fluoride crystal. Generally the limiting velocity appears to be the elastic wave speed (the speed of sound) for shear stress, but the relationship does not appear to be simple (McClintock and Argon, 1966, Fig. 4.20).

As temperature decreases the velocity at which dislocations move is reduced. As strain rate increases the time available for dislocations to move is reduced. In either case the pattern usual at low strain rate and room temperature may not have time to develop, and another pattern takes its place. In particular there is a phenomenon known as *twinning* in which a crystal deforms along a line in such a way that the lattice pattern on one side of the line is a mirror image of the pattern on the other side of the line. Twinning tends to become more common as strain rate is increased or temperature is reduced (McClintock and Argon, 1966, p. 134).

4

Fracture

4.1 INTRODUCTION

4.1.1 General

(a) Traditional impact tests

A number of traditional impact tests have been in use for an appreciable time. They preceded the extensive study of the properties of materials at high strain rates and the development of modern theories of brittle fracture and so they are basically empirical. Rates of strain are not usually defined and crack propagation speeds are usually ignored. No consistent correlations exist among different tests, but they are widely used for rank order comparisons of fracture resistance, for establishing ductile–brittle transition temperatures and for quality control. As long as the overall behaviour of the material is well understood they are useful, but they can give misleading results if taken out of context and should be used with caution.

With some high strength steels, slow bend tests on notched specimens give a decrease in energy to cause fracture as temperature increases but Charpy impact tests give an increase in energy. Similar behaviour can occur with titanium alloys, as is shown in Fig. 4.1 (Hartbower, 1969).

Instrumented Charpy tests have shown that the energy to break the specimen can be divided into two components – that needed to start a crack, and that needed to propagate it. Figure 4.2 shows how the two energies vary over the temperature transition range. The strain rate at the notch root in a standard Charpy test is around $10^2 \, s^{-1}$ (Fearnhough, 1973).

(b) Factors affecting fracture type

There are two main groups of factors which affect the susceptibility of a material to brittle fracture. These are the intrinsic properties of the material itself and the conditions under which it is loaded. There is some overlap between the groups. For instance in a crystalline material containing a large number of grains the average grain size is an intrinsic material property, but in small-scale investigations involving a single grain its size and shape are aspects of the loading conditions.

The intrinsic properties can be varied to a considerable extent although

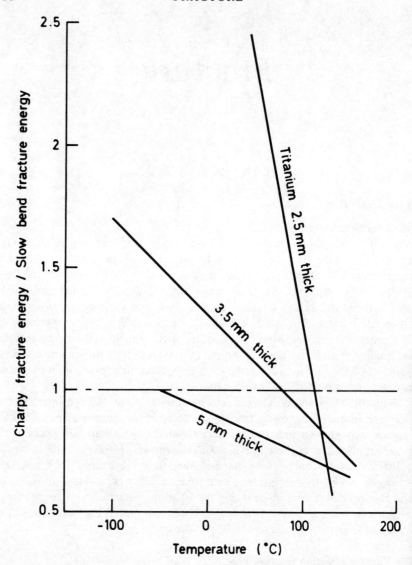

Figure 4.1 Fracture energy in Charpy impact and slow bend tests: variation with temperature.

often an improvement in one property is achieved at the expense of another. Changes in intrinsic properties can also occur with time. Corrosion or fatigue can produce fine cracks in an originally ductile material leading to its failing in a brittle manner as the cracks suddenly spread.

The loading factors can be divided into two different types: geometric,

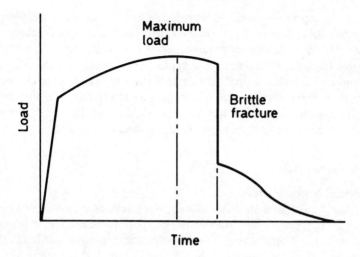

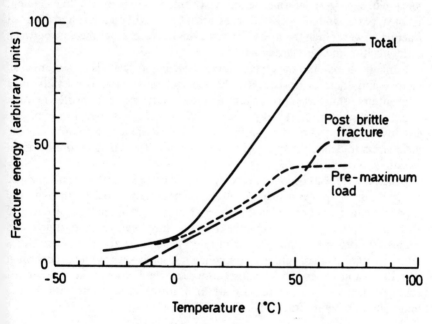

Figure 4.2 Behaviour in instrumented Charpy test.

covering size and shape; and thermodynamic, covering temperature and strain rate. These can be studied separately, but often no attempt is made to isolate the different aspects. Stress concentrations and increasing size increase the tendency to brittle fracture. Decreasing the temperature and increasing the strain rate tend to produce more brittle behaviour, but detailed relationships can be complex and sometimes the two parameters can have opposing effects. Some materials are more sensitive than others. Pure iron usually has a body centred cubic lattice and becomes more brittle with decreasing temperature. Pure aluminium has a face centred cubic lattice and is not sensitive to changes in temperature. Pure titanium has a hexagonal close packed lattice; it is sensitive to temperature changes but not as much as iron. The behaviour of metal alloys is more complex but the general patterns still stand.

4.1.2 Basic relationships

(a) Ductile and brittle fracture

If a piece of solid material is strained sufficiently it fractures, breaking into two or more separate pieces. There are two basic types of fracture – ductile and brittle. A completely brittle material shows no plastic behaviour but fractures suddenly from a completely elastic state. A completely ductile material in a tension or shear test deforms continuously until it is reduced to a fine filament before it parts. Under hydrostatic pressure it behaves like a liquid and does not fracture however high the stress. In most cases fracture is intermediate between purely brittle and purely ductile.

For a given material the type of fracture occurring depends on a number of factors and there is not necessarily a correlation between ductility in a conventional static test at room temperature and resistance to brittle fracture. A separate parameter called *fracture toughness* (K_{IC}) is used to define the conditions for the onset of brittle fracture. Its use is analogous to the use of the yield stress to define the onset of plastic flow but it is a much more complex parameter (see Section 4.3.1(a)).

Brittle fracture has been studied very extensively because it is the main cause of failure in components made from brittle materials. It occurs through the rapid propagation of cracks in the material, at the maximum load carried and with little prior warning. Ductile fracture has been studied much less extensively because, although it is of interest in the study of materials, it is seldom a main cause of failure in components. Usually such failure occurs when a limiting value of permanent deformation is reached or when the load carried starts to decrease with increasing strain. Both of these occur with ductile materials well before fracture.

(b) Fracture toughness

Fracture toughness is a measure of the resistance of a material to the propagation of the cracks which lead to brittle fracture. It is measured on static test specimens of standard proportions. A low stress fatigue crack is produced in the specimen; it is assumed that this is the sharpest crack which can occur in

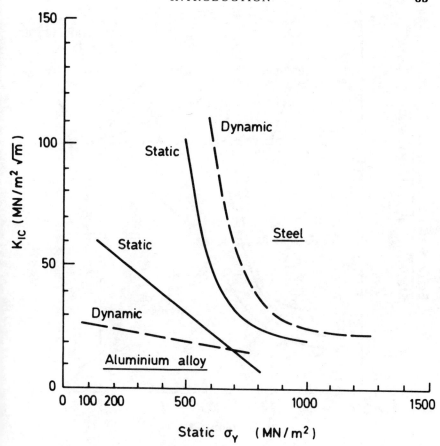

Figure 4.3 Variation in fracture toughness K_{IC} with static 0.2% yield stress σ_y at room temperature.

the material, so that it is a repeatable worst case and needs no further specification. The specimen is then loaded statically until it fractures and the fracture toughness is derived from the load–deflection measurements using standard procedures and tables.

Generally fracture toughness decreases with increasing yield stress. This can be seen in Fig. 4.3 (Andrzejewski *et al.*, 1981, Table 1; Krabiell and Dahl, 1981, Fig. 9). It also decreases with decreasing temperature in those materials where σ_y is temperature dependent. Behaviour is not simple, as can be seen from Fig. 4.4. With aluminium alloy there is no temperature effect. With three structural steels, fracture toughness either falls steadily as temperature is reduced below room temperature or first increases and then decreases. With a

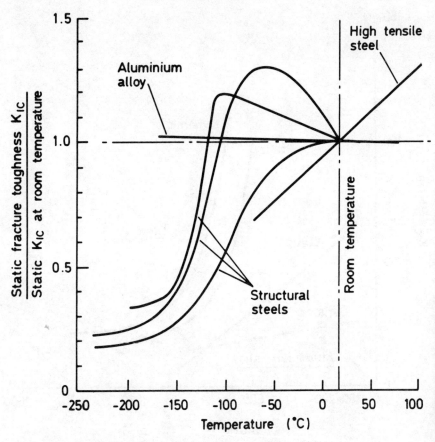

Figure 4.4 Variation in static fracture toughness with temperature.

maraging steel fracture toughness falls steadily below 0 °C but remains constant above that, and with a high tensile steel it decreases steadily with decrease in temperature (American Society of Metals, 1975, Vol. 10; Krabiell and Dahl, 1981, Fig. 9).

Fracture toughness is a complicated parameter and the analysis of any particular case can become very involved. Many representative static cases have been worked out and the results published in manuals, so it is usually sufficient to refer to these when considering the effects of changes in static loading conditions.

4.2 DUCTILE FRACTURE

Fracture is the end point of any of the tests used for measuring yield stress, ultimate tensile stress and plastic behaviour. Continuous measurement of load can be made up until fracture and the residual permanent deformation can be measured on the broken specimen. *Ductility* is generally defined by some simple function of the permanent residual strain, and empirical relationships have been established between these measures of ductility and behaviour in service.

The behaviour of specimens in static tests can be modified by varying the temperature and by introducing stress-raising notches. The elastic–plastic stress distributions around notches are easier to analyse than those around cracks. Static tests of notched specimens have been used to study behaviour at and close to fracture, including size effects, notch sensitivity, temperature effects and correlations between macroscopic and microscopic behaviour.

In general a material can fracture in either a ductile or a brittle manner and adopts the type which needs the lower stress. The stress to cause brittle fracture

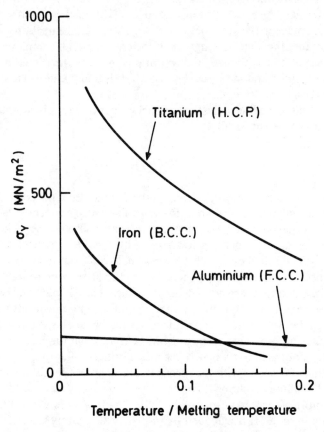

Figure 4.5 Variation in static yield stress with temperature for pure metals of three different lattice types.

is usually insensitive to temperature, as shown in Fig. 3.3. However, in body centred cubic and hexagonal close packed metals the yield stress rises as the temperature falls, as can be seen in Fig. 4.5, and so these metals tend to change from ductile to brittle behaviour with decreasing temperature (Wigley, 1971, p. 37). Sometimes this transition can occur over a narrow temperature range. The effects of notches and temperature are additive, and so a notch can raise the transition temperature.

4.3 FRACTURE MECHANICS

4.3.1 Static

(a) Linear elastic fracture mechanics
A tensile stress cannot cross a crack; thus, if a transverse crack is present in a tensile specimen, the material along the edges of the crack is unstressed. At the ends of the crack, on the other hand, very high stress concentrations occur and these can cause the crack to lengthen. If the crack is small compared with the dimensions of the specimen, the variations in stress are confined to its immediate vicinity and the specimen as a whole is under uniform tensile stress as if no crack were present.

The analysis of the linear elastic stress field around a crack is complex, but in all cases it can be shown that

$$\sigma_{ij} = \frac{K}{\sqrt{(2\pi r)}}$$

where σ_{ij} is the stress in any direction, r is the distance from the crack tip and K is a constant called the *stress intensity factor*. The use of a constant value for K is an approximation which becomes less accurate as r increases. The units of K are rather curious, being stress $\times$ (length)$^{1/2}$, which reduces to force $\times$ (length)$^{-3/2}$. This makes it difficult to visualize K, but it is an extremely useful concept. In real materials there is a small plastic zone round the crack tip in which conditions are not amenable to linear elastic analysis. However, it is assumed that, for any given material, there is an unknown but consistent relationship between crack propagation and the linear elastic stress intensity factor immediately outside the plastic zone. In particular the crack will start to propagate spontaneously when K reaches a critical value K_{crit}.

The value of K varies with the specimen dimensions, the loading mode and the overall loading conditions. The crack tip can be regarded as being loaded in one or more of three basic modes. In mode I loading the crack is pulled apart by tension; in mode II it is loaded by in-plane shear along the crack; and in mode III it is loaded by anti-plane shear across the crack. For each mode there is a different value of K, and these are called K_I, K_{II} and K_{III}, respectively. They can be treated separately or in combination, but K_1 is used much more often than the other two. There are two limiting conditions for overall loading. In the first, which is known as *soft loading*, the force on the specimen remains constant so that once a crack has started to propagate spontaneously it keeps on going. A typical case is a specimen loaded by a weight hanging from it. In the second, which is known as *hard loading*, the overall displacement of the specimen remains constant and the amount of energy available is limited to the elastic strain energy stored in the specimen when the crack started to propagate spontaneously. In this case K decreases as the crack propagates. A typical case is a specimen loaded in a stiff testing machine with fixed loading grips.

In thin specimens the stress system is plane stress but in thick specimens it is plane strain because lateral contraction is prevented. In many cases the latter gives the lowest value of K_{crit}. Usually mode I loading is used, giving plane strain fracture toughness K_{IC}. The fracture toughness can be regarded as an inherent property of the material.

(b) Yielding fracture mechanics
With ductile materials a crack may occur but not become unstable so that it only grows if energy is continuously supplied from outside. In this case there can be considerable scatter in test results. If the components to be used in service are not thick enough for plane strain cracking, an appropriate value of K_1 higher than K_{IC} can be used. It may also be desirable to estimate K_{IC} from tests on relatively thin specimens where plane strain is not reached. In either case it is necessary to know how values of K_1 under different conditions are related to each other. Strictly the conditions leading to crack instability can no longer be summarized by a single parameter, but it has been found possible to use single parameters over fairly wide ranges of behaviour. In general, if crack instability is reached before general yielding of the specimen, K_{IC} is used. If it occurs after general yielding, a modified approach is needed.

The energy in a specimen containing a crack can be studied over a wide range of conditions by using the *path-independent integral J*. The analysis can deal with non-linear elasticity; it is assumed that it can also be used to cover specimens which have yielded overall so that their behaviour, though largely elastic, is partly plastic. This cannot be justified analytically but good results are obtained in practice as long as the fractures in yielded specimens are broadly similar to those in elastic specimens. The analysis cannot deal with the fully plastic zone at the crack tip.

In two dimensions the integral can be calculated along any arbitrary line starting on one side of the crack and running round the crack tip to the other side. It is derived by considering the forces acting on any small length of the line and the elastic displacements caused by them when the crack extends by an infinitesimal amount. The forces do work in causing the displacements and the resulting change in the elastic stress field changes the elastic strain energy stored in the material within the line. If these changes are integrated along the whole length of the line the resulting integral J is equal to the total elastic work done by the forces plus the total change in the elastic strain energy of the material within the line.

Because J can be calculated along any line it is no longer subject to the limitations of the approximation for K which becomes less accurate as r increases. It is still necessary to assume a consistent relationship between J at the edge of the fully plastic zone round the crack tip and crack propagation within the zone, so that the crack becomes unstable at a critical value J_{crit} at this boundary. Experimental values of J_{crit} can be derived from the load–deflection characteristics of a cracked specimen in the same way as K_{crit}. It has been found that for mode I loading J_{IC} adequately defines the start of crack propagation in specimens of different geometries. However, the method is still rather controversial.

J is wider in its application than K but it is more complex and, in the region where linear elastic fracture mechanics can be used, it has no obvious advantage. It can be extended to a surface independent integral in three dimensions.

4.3.2 Dynamic

Dynamic fracture mechanics is a logical extension of static fracture mechanics, but analysis can become extremely complex and behaviour tends to be treated semi-empirically using static behaviour as a basis. In addition to the analytical complexity, dynamic test results can be difficult to interpret because inertia effects, vibrations and stress wave interactions can all be present and can cause unexpected results.

There are three distinct aspects of dynamic crack behaviour. These are the initiation of instability in a stationary crack when the stress intensity is increasing rapidly; the propagation of a moving crack; and crack arrest. They will be discussed in turn.

(a) Crack initiation

There is no simple universal correlation between the strain rate $\dot{\varepsilon}$ and the stress intensity rate $\dot{K}$ because K is a function of the distance from the crack tip as well as of the stress. They can be related only at specific locations and the relationship varies with the location chosen. Usually the interface between the plastic and elastic regions at the crack tip is used because the stress there is the yield stress σ_y or a function of the yield stress. This interface can be located on the calculated elastic stress field but it is necessary to know both $\dot{\varepsilon}$ and the appropriate dynamic value of σ_y, and so both variables need to be considered simultaneously. The relationship between $\dot{\varepsilon}$ and σ_y is taken from dynamic tensile tests.

The position of the interface defines the size of the plastic zone. This varies with the material, the temperature, the geometry of the specimen and the strain rate, and so it is not possible to give a single consistent relationship between K and ε at the interface. Estimates of the relationship for steel appear broadly to give $\dot{\varepsilon} = 1\,\mathrm{s}^{-1}$, equivalent to $\dot{K} = 10^3$ to $10^6\,\mathrm{MPa\,m^{1/2}\,s^{-1}}$.

Dynamic plane strain fracture toughness is written K_{ID}. There is not a simple correlation with K_{IC}. Figure 4.6 shows behaviour at room temperature for crack initiation. As $\dot{K}$ increases, K_{ID} increases slightly for structural steels, decreases slightly for rail steel and aluminium alloy, and decreases and then increases for a high tensile cast steel. Detailed behaviour can also vary markedly for different heats or heat treatments of a given metal (Krabiell and Dahl, 1981, Figs 5, 7, 8; Andrzejewski et al., 1981, Fig. 5; American Society of Metals, 1975, Fig. 10; Christopher et al., 1981, Fig. 2).

Under severe shock loading it appears that individual cracks do not have time to propagate but that large numbers of small cracks are formed. This

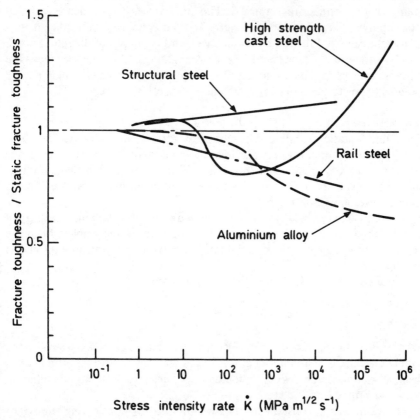

Figure 4.6 Variation in fracture toughness K_{IC} with stress intensity rate $\dot{K}$ at room temperature.

reduces K_{ID} to two orders of magnitude below K_{IC} for pure iron and one order of magnitude below K_{IC} for beryllium.

(b) Crack propagation

Resistance to crack extension
Once a crack has begun to propagate spontaneously its further extension depends on three factors: the elasticity of the loading system, the specimen dimensions and the material properties. Whether the loading system is sufficiently soft to keep the crack running depends in part on the other two.

In general, for metals, there is more variation in the stress intensity needed to keep a crack running than there is in the stress intensity to initiate it, and metals which have similar values of K_{IC} or K_{ID} can be markedly different in their resistance to subsequent crack propagation. One well-known method of studying this is the dynamic tear test developed by the US Navy. The test is intended for engineering use and not for basic study of fracture mechanics, so many of the results are empirical. The standard test specimen is a beam containing a sharp notch at mid length. Test pieces are generally taken from a plate and can be of the full plate thickness, and the depth of the initial notch can be varied to alter the length of the propagating crack. The specimens are broken by a pendulum or falling weight which strikes the specimen at mid length, opposite the notch, and the energy absorbed is measured.

Crack velocity
Even under static loading cracks can propagate at high velocity. The static energy balance, for constant stress loading, gives the strain energy release rate increasing as the crack lengthens. Thus, if the formation of new surfaces absorbs energy at a constant rate, there is an increasing amount of energy available to accelerate the crack.

The static energy balance can be extended to include the kinetic energy of the material set in motion as the crack tip passes. In its simplest form the analysis ignores changes in material properties with strain rate and the effects of stress waves.

For a constant tensile stress this analysis gives

$$v = \text{constant} \times C \sqrt{\left(1 - \frac{a_0}{a}\right)}$$

where v is the crack velocity, C the velocity of a direct stress wave in a rod, a the crack length and a_0 the initial crack length when spontaneous propagation began. As the crack length a increases v tends to a terminal value of $v_T = \text{constant} \times C$.

The value of the constant term can be derived by considering the strain energy release rate. In its simplest form the analysis assumes a circle of radius r

round the crack tip. Strain energy released from material within the circle has time to reach the crack tip and influence its behaviour. Strain energy released from material outside the circle does not. It takes the unloading elastic wave a time $t = r/C$ to travel from the periphery of the circle to its centre and, in this time, the crack can lengthen by $\delta a = v_T t$. Eliminating t gives $r/\delta a = C/v_T$, and restricting the energy balance within the circle gives $v_T = 0.38\,C$.

Although the allowance for the effect of stress wave velocity is crude, the result is borne out by experiment. Maximum crack speeds observed in brittle materials range from about $0.2\,C$ for steels to about $0.4\,C$ for quartz. More sophisticated analyses have been made and, in some of these, the velocity of shear waves or surface waves is taken as the reference velocity. This can be confusing.

With constant displacement loading the energy available to propagate the crack is the strain energy in the specimen at the onset of crack instability. This varies with the length of the initial crack relative to the size of the specimen. A short initial crack in a large specimen behaves in much the same way as a crack under constant stress loading. A long initial crack in a small specimen propagates spontaneously for only a short distance and then stops. Cracks of intermediate length accelerate to a maximum velocity lower than v_T and then, if they have not already reached the specimen edge, they stop.

In low speed crack propagation behaviour is isothermal as the heat generated in plastic deformation has time to escape. In high speed propagation this heat has no time to escape, behaviour is largely adiabatic and the resulting rise in temperature at the crack tip increases the value of K_I needed to keep the crack running. At impact loading rates behaviour seems invariably to be adiabatic, and Fig. 4.7 shows values of K_I against crack velocity for mild steel and PMMA (Eftis and Krafft, 1965; Radon and Fitzpatrick, 1973).

The appearance of crack surfaces changes with crack velocity. With glass the surface gets rougher as the crack speed increases but with steel and polymers it gets smoother.

Crack paths
In a simple tension specimen an initially existing crack is likely to lie at an angle to the line of action of the applied stress, so that it is loaded partly in mode I, with a tensile force pulling the crack apart, and partly in mode II, with a shear force acting along the crack. Static analysis of a homogeneous isotropic material indicates that mode I loading is stable but mode II loading is not and tends to disappear as the crack propagates.

As a result the crack curves until it is lying at right angles to the line of action of the applied stress and it is loaded entirely in mode I. If the material is not homogeneous it may contain preferred planes of cleavage and the crack will tend to follow these instead.

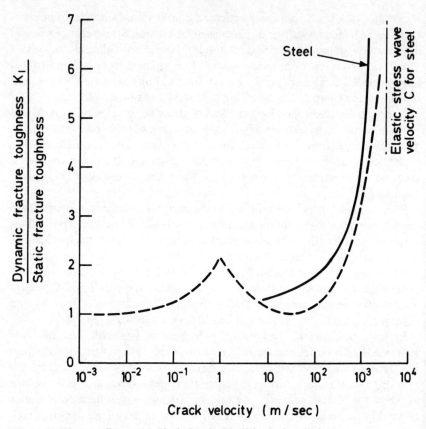

Figure 4.7 Variation in K_I with crack velocity.

In a simple compression specimen a crack which is at right angles to the line of action of the applied stress is held shut and does not propagate. Static analysis indicates a high tensile stress concentration on the side of an inclined crack, a small distance from the tip. As the crack propagates it curves towards this stress concentration until it lies parallel to the line of action of the applied stress and then it stops. A single crack is thus unlikely to cause fracture but a number of cracks can run together to do so.

If a crack is loaded by a combination of mode I and mode III loading, the plane of the crack rotates about its longitudinal axis as it propagates, forming a helix. In real materials it tends to do this in a series of short, flat steps rather than in a continuous curve, giving a hackled effect. If all three modes of loading are present, crack paths become complex.

In a fast moving crack the theoretical linearly elastic stress field changes. In static mode I loading the maximum tensile stress is directly ahead of the crack tip but in high speed loading the maximum tensile stress occurs at two positions symmetrically disposed on either side of the crack tip, leading to a tendency for the crack to branch in two.

Crack branching occurs mainly in very brittle solids and so only a limited range of structural materials exhibit branching, but the likelihood can be increased by corrosion. Branching appears to absorb a large amount of energy and so it can slow down the crack. High crack velocity seems to be a necessary condition but not, by itself, a sufficient one. There also needs to be a stress intensity K_B which generally seems to be about 2 to 3.5 times K_{IC} for glass and 5 to 10 times K_{IC} for hard steel (Congleton, 1973).

(c) Crack arrest

A moving crack can arrest spontaneously. There are two main ways in which this can happen: either the strain energy release rate falls below that needed to keep the crack propagating, or conditions change in such a way that further propagation of the crack requires a marked increase in the strain energy release rate. The first of these occurs under stiff loading conditions in which there is only a limited amount of strain energy available and, when this is used up, the crack stops. The stress intensity at which this occurs is called the *crack arrest intensity* K_a. This is not necessarily a unique value, and it has been suggested that it ought to combine an inherent material property with a size and loading condition factor for the amount of kinetic energy present in the moving crack and available to keep the crack extending as it slows down.

The second type of spontaneous crack arrest occurs in temperature-sensitive materials when a temperature gradient is present. A crack which starts in the colder, more brittle region can arrest when it reaches the warmer, more ductile region. The temperature at which this occurs is called the *crack arrest temperature*.

Tests for crack arrest are usually carried out under stiff loading. It is possible for cracks to fail to arrest in service, under conditions which are safe in such tests, if large amounts of strain energy or kinetic energy are present.

5

Rigid body motion

5.1 INTRODUCTION

5.1.1 General

(a) Scope of chapter

Analysis of the impact of rigid bodies is simpler than other forms of impact analysis because any deformations of the bodies are ignored. It illustrates overall patterns of behaviour in the simplest possible way and is the only form of impact analysis generally considered in textbooks on dynamics. The behaviour of non-rigid bodies can often be divided into overall motion, which can be treated by rigid body analysis, and deformations, which can be dealt with separately.

There are three basic types of analysis which are widely used. Each of these has its advantages and disadvantages and each is more widely used in some disciplines than in others. As a result an impact may be analysed by a particular method sometimes because it is the best method but often because it is the method with which the writer is most familiar. This can make it difficult for the non-expert who is not familiar with the particular method and who may not understand why it has been used.

Comprehensive study of any of these methods is beyond the scope of this book and there are excellent textbooks dealing with all of them, though not in the context of impact. This chapter aims to provide a very simple overall appreciation of the principles involved and of the notation used, so that the reader can get a general impression of the gist of published papers even if he cannot follow the analysis in detail.

(b) Types of analysis

Analysis of behaviour during an impact involves the setting up and solving of equations of motion. In much impact analysis fairly crude approximations are used because loading conditions are known only approximately and material properties are poorly understood. There are three basic ways of dealing with the equations of motion:

(1) Force balances can be established, leading to differential equations which

are often readily soluble either analytically or numerically. This method is widely used but the differential equations can become complicated.

(2) Energy balances can be established, leading to integral equations which are usually more difficult to solve than differential equations. However, energy is often an easier parameter to deal with than force. In Lagrangian dynamics energy balances are converted to force balances which are then used to establish differential equations. These are often less complicated than the equations found directly from force balances, although they are more difficult to derive.

(3) Impulse–momentum balances can be established, and in its simplest form the method is very useful and simple. More detailed analysis leads to convolution equations which can be difficult to solve.

5.1.2 Displacement, velocity and acceleration

Displacement, velocity and acceleration are interrelated, and sometimes the critical step in an analysis consists in changing an expression in terms of one into an equivalent expression in terms of another. A number of equivalents is given in Table 5.1.

Velocity is speed in a specified direction. Speed and velocity are relative and not absolute parameters and can only be specified by the rate at which the distance between two locations changes. There is no fundamental way of defining which location is moving, and there is no loss of generality if one of them is assumed to be stationary at any convenient time. The stationary location and the convenient time can be chosen to give the simplest possible analysis. Sometimes it is more convenient to treat a third location as stationary, for instance the position of the combined centre of mass of two moving bodies. There is always a change in velocity during an impact, and the

Table 5.1 Expressions for displacement, velocity and acceleration.

Displacement	Velocity	Acceleration
x	v	a
	$\dfrac{dx}{dt}$	$\dfrac{dv}{dt}$
		$v\dfrac{dv}{dx}$
		$\dfrac{d^2x}{dt^2}$

total velocity change is often used as the basic parameter defining impact severity. Strain rate, the reference parameter used in the study of material properties under impact, is a normalized measure of velocity.

Displacement is obtained by integrating velocity with respect to time. Like velocity it is a relative parameter and can only be specified by the distance between two locations. Acceleration is the rate of change of velocity and can be obtained by differentiating velocity with respect to time. Unlike the other two it is an absolute parameter which can be specified at a single location with no external reference. For this reason it is often convenient to use the maximum or average acceleration to define the severity of an impact as an alternative to the total velocity change.

In theoretical analysis, using well-behaved mathematical functions, both differentiation and integration are exact; however, with empirical records, differentiation is not reliable and can produce cumulative errors as large as the expected answer. As a result detailed numerical differentiation of empirical results is avoided whenever possible. Integration, despite some problems, is much more reliable with empirical data. The practical consequence of this is that, although displacements and velocities often can be derived from measured accelerations, it is difficult to derive accelerations from measured velocities or velocities from measured displacements. Accelerations derived from measured displacements by a double differentiation may be completely unreliable.

5.2 SIMPLE ANALYSIS

5.2.1 Linear motion in one dimension

(a) Force balances

Force, mass and acceleration
Newton's second law of motion can be written

$$F = Ma \tag{5.1}$$

where F is force, M is mass and a is acceleration. The three parameters are interrelated and need to be specified in terms of each other.

At the earth's surface a convenient constant force is provided by gravity which accelerates a body downwards if it is free to fall. The weight of the body is the upwards force needed to prevent it from falling. Acceleration in free fall can be measured directly, and weight can be measured independently by the deflection of a calibrated spring. Thus if the relationship is rewritten

$$W = Mg \tag{5.2}$$

where W is weight and g the acceleration due to gravity, then $M = W/g$. The units of mass are kilograms (kg), of acceleration are metres per second per second ($\mathrm{m\,s^{-2}}$) and of weight are newtons ($\mathrm{N = m\,kg\,s^{-2}}$). Often accelerations are expressed as multiples of g which provides a simple intuitively appreciated reference value.

Forces usually vary during an impact but masses usually remain constant, the main exception being the propagation of a stress wave when the mass in motion changes with time. In all the analyses considered here, except stress wave analysis, masses are assumed to remain constant. Force and mass can then be used interchangeably through Equation 5.1.

Constant acceleration
The simplest case is when the acceleration remains constant, giving the well-known relationships

$$v = u + at \tag{5.3}$$

$$v^2 = u^2 + 2ax \tag{5.4}$$

$$x = ut + \tfrac{1}{2}at^2 \tag{5.5}$$

where v is final velocity, u is initial velocity, x is displacement and t is the time from the start of the acceleration.

Acceleration pulses
During an impact, acceleration occurs as a pulse which lasts for a short time. Often pulse shapes are not known or are too complicated for simple analysis, and simple approximate shapes are used instead. The shapes chosen depend on the type of impact being analysed, and so published work concentrates on different simple shapes in different areas and there is no single simple pulse of universal application.

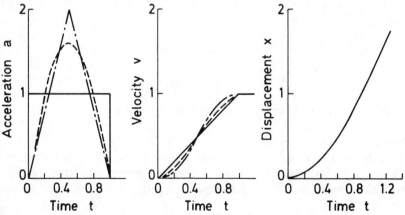

Figure 5.1 Velocity and displacement for simple acceleration pulses (arbitrary units).

Three simple symmetrical acceleration–time pulses are shown in Fig. 5.1 together with the resulting velocities and displacements. Note how in this case different shapes of acceleration pulse produce the same shape of displacement–time curve, illustrating how difficult it is to derive an acceleration from a displacement. It may be impossible to distinguish between alternative theories based on force–time relationships if only displacements can be measured.

(b) Energy balances

Work, virtual work and d'Alembert's principle
Work is defined as force multiplied by displacement, so that

$$W = \int F \, dx \tag{5.6}$$

where W is work, F is force and x is displacement. The units of work are newton metres (N m); to avoid confusion with moments, which have the same units but mean something different, the unit of work is given the distinctive name of the joule (J). Moments are discussed in Section 5.2.2.

It is often convenient, in setting up a mathematical model, to consider the work which a force would do if it were to move through a small displacement δx even though no such displacement occurs. This is known as *virtual work* ($\delta W = F \delta x$), and the hypothetical displacement is known as the *virtual displacement*.

D'Alembert's principle relates work to Equation 5.1 by multiplying each side by a small displacement δx, giving, in the limit,

$$F \, dx = Ma \, dx$$

If the acceleration is replaced by the appropriate velocity relationship from Table 5.1, this becomes

$$F \, dx = Mv \frac{dv}{dx} dx$$

$$= Mv \, dv \tag{5.7}$$

which can be used for both actual and virtual work.

Potential and kinetic energy
Energy is the capacity to do work, and the units of energy are the same as those of work. It can take many forms, such as chemical energy or electrical energy, but the two forms mainly used in impact analysis are potential and kinetic energy.

The work done equals the energy released, but usually it is more convenient to relate the work done to the energy remaining. For potential energy this relationship can be expressed as

$$\delta W = -\delta V$$

where δW is a small increase in the work done and $-\delta V$ is the corresponding small decrease in the potential energy remaining. It follows from Equation 5.6 that $F\delta x = -\delta V$ or, in the limit,

$$F = -\frac{dV}{dx} \tag{5.8}$$

Kinetic energy is possessed by a mass as a result of its velocity, and it equals the work needed to bring the mass up to this velocity or done by the mass if it loses its velocity. The relationship can be derived from d'Alembert's principle by integrating both sides of Equation 5.7 to give

$$\int_{x_1}^{x_2} F \, dx = \int_{u}^{v} Mv \, dv$$

where u is velocity at displacement x_1, and v is velocity at displacement x_2. This reduces to

$$W = \tfrac{1}{2}M(v^2 - u^2) \tag{5.9}$$

The overall kinetic energy is taken as that relative to the body whilst it is stationary, in which case $u = 0$ and overall kinetic energy is

$$T = \tfrac{1}{2}Mv^2 \tag{5.10}$$

As with potential energy, it is convenient to relate the work done to the kinetic energy remaining in the moving body, giving $\delta W = -\delta T$ or, in the limit,

$$F = -\frac{dT}{dx} \tag{5.11}$$

Note that, if Equation 5.11 is written as $F \, dx = -dT$ and integrated, the result is

$$\int_{x_1}^{x_2} F \, dx = -\int_{T_1}^{T_2} dT$$

where T_1 is the kinetic energy of the mass at displacement x_1, and T_2 is the kinetic energy at displacement x_2. This gives

$$W = -(T_2 - T_1)$$

$$= \frac{M}{2}(u^2 - v^2) \tag{5.12}$$

Equations 5.12 and 5.9 are the same except that the signs are reversed; this is because Equation 5.9 states the total kinetic energy whereas Equation 5.12 states the reduction in the kinetic energy remaining as work is done. This can be confusing, but common sense usually indicates which sign to use.

Conservative and non-conservative systems
If there are only potential and kinetic energy present and there are no energy losses, then the total energy in the system is

$$E = V + T \tag{5.13}$$

The system is said to be conservative and E remains constant. If both sides of Equation 5.13 are differentiated with respect to displacement, then

$$0 = \frac{dV}{dx} + \frac{dT}{dx}$$

or

$$dV = -dT \tag{5.14}$$

More generally, if the work done by a force depends only on its initial and final positions, regardless of the path it follows between them, the force is conservative. A system which contains only conservative forces is a conservative system; no energy losses occur in it, and transfers of energy from one form to another are completely reversible.

In non-conservative systems losses occur, and so Equation 5.13 needs to be modified to include these. Almost all real systems are non-conservative to a greater or lesser degree but often it is convenient to analyse them as if they were conservative. This gives a simpler analysis and often a clearer insight into patterns of behaviour. Corrections for the losses can be incorporated later.

If the energy lost is regarded as the total work done W up to the time of the analysis then, at any time,

$$E_0 - W = V + T \tag{5.15}$$

where E_0 is the initial total energy. Alternatively, if the losses are made good by work done from outside, the total energy remains constant and

$$E = V + T + W$$

In either case, if both sides are differentiated with respect to displacement,

$$\frac{dW}{dx} = -\left(\frac{dV}{dx} + \frac{dT}{dx}\right)$$

or

$$F = -\left(\frac{dV}{dx} + \frac{dT}{dx}\right) \tag{5.16}$$

This is a more general form of Equations 5.8 and 5.11. Note that F has the units of force, but it is not quite the same force as defined in Equation 5.1.

Force–displacement pulses
Sometimes a pulse is known in terms of force–displacement rather than force–time. This makes the analysis much less tractable. If the force is constant, Equation 5.5 can be rewritten to give

$$\tfrac{1}{2}at^2 + ut - x = 0$$

This leads to

$$t = \frac{-u \pm \sqrt{(u^2 - 2ax)^{1/2}}}{a}$$

which can be solved to give the variation in displacement with time.
   If the force is not constant, Equation 5.12 can be used. Thus

$$W = \frac{M}{2}(u^2 - v^2)$$

giving

$$\frac{dx}{dt} = v = \pm \sqrt{\left(u^2 - \frac{2W}{M}\right)}$$

or

$$dt = \frac{dx}{\pm \sqrt{[u^2 - (2W/M)]}}$$

Integrating both sides gives

$$t = \pm \int_{x_1}^{x_2} \frac{dx}{\sqrt{[u^2 - (2W/M)]}} \qquad (5.17)$$

This is a standard integral equation which can be solved by standard procedures to give the variation in displacement with time.
   Variation in velocity with either displacement or time is found by differentiating the displacement–time curve.

Lagrangian dynamics
It is possible to use energy balances to produce differential equations in terms of forces by using Lagrangian dynamics. Apart from avoiding integral equations, this approach has two advantages: equations of motion can be stated in a common form, regardless of the system of co-ordinates used; and forces which do no work can be eliminated from the analysis. The disadvantage is that the original setting up of the equations can be rather difficult.

The method is useful in the analysis of complex systems in three dimensions but, for illustration, it is discussed in terms of a single mass moving in one dimension.

Equation 5.10, which relates kinetic energy to velocity, gives

$$\frac{dT}{dv} = \frac{d(\frac{1}{2}Mv^2)}{dv}$$

$$= Mv \tag{5.18}$$

as long as M remains constant and no constant term is present. Now

$$Ma = M\frac{dv}{dt}$$

$$= \frac{d(Mv)}{dt}$$

as long as M remains constant. Substituting from Equation 5.18,

$$Ma = \frac{d(dT/dv)}{dt} \tag{5.19}$$

Adding Equation 5.1, which relates force to acceleration, to Equation 5.11, which relates force to change in kinetic energy with displacement, gives

$$F = Ma - \frac{dT}{dx}$$

$$= \frac{d(dT/dv)}{dt} - \frac{dT}{dx} \tag{5.20}$$

In a conservative system the force producing change in kinetic energy T is the same as the force producing change in potential energy V, which is (from Equation 5.8)

$$F = -\frac{dV}{dx}$$

Eliminating F gives

$$\frac{d(dT/dv)}{dt} - \frac{d(T-V)}{dx} = 0 \tag{5.21}$$

This is the Lagrangian equation of motion for a simple conservative system.

In a non-conservative system energy losses are present, so that either the total energy decreases or work needs to be done from outside the system to keep the total energy constant. In either case the Lagrangian equation of motion can be shown to be

$$\frac{d(dT/dv)}{dt} - \frac{d(T-V)}{dx} = \frac{W}{dx}$$

where W is the work done from outside the system or lost within the system. All of the terms are in units of force, and the equation is usually written

$$\frac{d(dT/dv)}{dt} - \frac{d(T-V)}{dx} = F \qquad (5.22)$$

In this context F is called a generalized force because it has the dimensions of force, but it can take a number of different forms, most of which are not forces in the simple sense.

(c) Impulse–momentum balances

Impulse and momentum
Impulse is defined as force multiplied by time, so that

$$I = \int F \, dt \qquad (5.23)$$

where I is impulse, F is force and t is time. The units of impulse are Newton seconds.

Momentum is defined as mass multiplied by velocity. It has the same units as impulse, and the two can be related by using Equation 5.1:

$$F = M\frac{dv}{dt}$$

so that

$$F \, dt = M \, dv$$

Integrating both sides gives

$$\int_{t_1}^{t_2} F \, dt = \int_{u}^{v} M \, dv$$

where u is the velocity at t_1 and v is the velocity at t_2. This gives

$$I = M(v - u) \qquad (5.24)$$

as long as M remains constant. If the initial velocity is zero then

$$I = Mv$$

or

$$v = \frac{I}{M} \qquad (5.25)$$

A useful feature of momentum is that it is always conserved. This is an empirical observation which appears to apply universally. In an impact the total momentum of any number of impacting bodies always remains constant, although the momentum of individual bodies within the system can vary and energy losses can occur.

Impulse–momentum balances are used mainly in situations in which the duration of the force is so short that no significant displacement occurs before it is over. The impulse can then be regarded as applying on instantaneous velocity change to the mass. In those circumstances Equation 5.22, the Lagrangian equation of motion, becomes

$$F = \frac{d(dT/dv)}{dt}$$

because there is no change in either potential or kinetic energy arising from displacement during the application of the force. This gives

$$\int_{t_1}^{t_2} F \, dt = \left(\frac{dT}{dv} \right)_u^v \tag{5.26}$$

and so the impulse equals the change in dT/dv. Note that F is a generalized force; thus the impulse is a generalized impulse and may take other forms than an impulse in the simple sense.

Mass ratios

If two bodies collide, the force between them and its duration may not be known but, because action and reaction are equal and opposite, $F_1 = -F_2$, where the subscripts denote the two different bodies. Using Equation 5.23 gives

$$I_1 = M_1(v_1 - u_1) = \int F_1 \, dt$$

$$I_2 = M_2(v_2 - u_2) = \int F_2 \, dt$$

Adding these gives

$$I_1 + I_2 = M_1(v_1 - u_1) + M_2(v_2 - u_2) = 0 \tag{5.27}$$

It is not possible to derive the velocities from this equation alone. However, the relationship between velocity change and mass can be established and is

$$\frac{M_2}{M_1} = \frac{v_1 - u_1}{-(v_2 - u_2)} \tag{5.28}$$

The velocity changes are in opposite directions and their magnitudes are proportional to the inverse ratio of the masses. In the limit, where one mass is infinite, all the velocity change occurs to the smaller mass.

Coefficient of restitution

Although momentum is always conserved there is usually a loss of energy on impact, and this can be dealt with by the use of an empirical parameter called the *coefficient of restitution*. This is always taken as positive and is given by the relative velocity of two masses after impact divided by their relative velocity before impact. It is often represented by the letter e. If there is no loss of relative velocity $e = 1$, and if all the relative velocity vanishes $e = 0$.

It was thought originally that e was a consistent property of a material, but is now accepted that e can vary greatly with input speed and loading conditions.

It is necessary to remember that two masses move closer together before impact and further apart after impact, so the sign of their relative velocity changes:

$$v_1 - v_2 = e[-(u_1 - u_2)]$$
$$= e(u_2 - u_1) \qquad (5.29)$$

This equation can be used in conjunction with Equation 5.28 to find velocities. There are seven variables – $M_1, M_2, u_1, u_2, v_1, v_2$ and e – but, because e is a function of the velocities and only the ratio of the masses is needed, only five variables are independent. If three of these are known the remaining two can be found from Equations 5.28 and 5.29.

In non-conservative systems, $e < 1$ and needs to be established empirically. It can be related to the loss in kinetic energy, and for illustration the case where $u_2 = 0$ is discussed. Equation 5.27 then gives

$$M_1(v_1 - u_1) + M_2(v_2) = 0$$

and Equation 5.29 gives

$$v_1 - v_2 = -eu_1$$

Substituting for v_2 gives

$$v_1 = \frac{M_1 - eM_2}{M_1 + M_2} u_1 \qquad (5.30)$$

and for v_1 gives

$$v_2 = \frac{(1 + e)M_1}{M_1 + M_2} u_1 \qquad (5.31)$$

The original kinetic energy is $T_0 = \frac{1}{2}M_1 u_1^2$ and the final kinetic energy is

$T_f = \frac{1}{2}(M_1v_1^2 + M_2v_2^2)$, giving

$$T_o - T_f = \frac{T_oM_1(1 - e^2)}{M_1 + M_2} \tag{5.32}$$

Acceleration pulses
The displacement resulting from a short duration impulse which occurs at time τ is

$$x = v(t - \tau)$$

where t is any time later than τ. In a longer duration impulse it may be necessary to allow for the displacement occurring during the impulse. This can be done by dividing the impulse into n smaller impulses each lasting such a short time that the displacement occurring during it is negligible. The total displacement is then

$$x = \sum_0^n v_n(t - \tau_n)$$

Using Equation 5.25 gives

$$x = \frac{1}{M}\sum_0^n I_n(t - \tau_n) \tag{5.33}$$

In the limit this gives

$$x = \frac{1}{M}\int_0^t F(t - \tau)\mathrm{d}\tau \tag{5.34}$$

This is a convolution integral, and the force F is said to be convolved with the time relationship. Standard procedures exist for the solution of such integrals, but the present case can be solved geometrically as shown in Fig. 5.2.

The small impulse I_n is the area of the strip $F\delta t$, and the expression $I_n(t - \tau_n)$ is the moment of this area about point t on the time axis. The total impulse I is the total area under the force–time curve, so that

$$\sum_0^n I_n(t - \tau_n) = I\bar{t}$$

where $\bar{t}$ is the distance between time t and the centroid of the area under the force–time curve. From Equation 5.33,

$$x = \frac{1}{M}\sum_0^n I_n(t - \tau_n) = \frac{I}{M}\bar{t} \tag{5.35}$$

The displacement at time t thus depends on the mass M, the total impulse I up to time t and the time between t and the centroid of the area under the force–

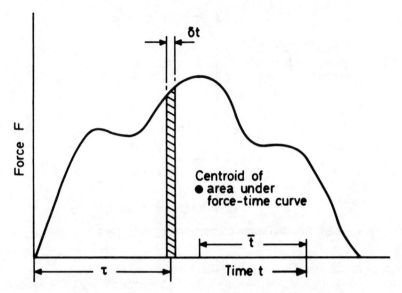

Figure 5.2 Geometric solution of convolution integral.

time curve up to t. In particular, if t is taken at the end of the pulse, and the force–time curve is symmetrical along the time axis, $\bar{t} = t/2$ and

$$x = \frac{I}{M}\frac{t}{2} = \frac{vt}{2}$$ (5.36)

5.2.2 Rotation about one axis

(a) Rotation only

Notation
Rotation about one axis involves motion in two dimensions, but it can be represented conveniently by a one-dimensional notation using the convention shown in Fig. 5.3. For simplicity, only rotation in a single plane is considered. The orientation of a plane is usually defined by the direction of the normal to it, and this is also the direction of the axis about which rotation occurs. The magnitude of the rotational motion is represented by length measured along this axis.

The positive direction is usually taken as the direction in which a right-hand screw would move if turned in the direction of the rotation. An alternative method is to hold the right hand with the thumb extended and the fingers lightly curled. If the fingers point in the direction of rotation, the thumb points in the positive direction along the axis.

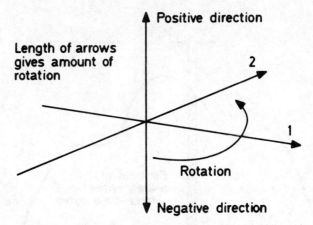

Figure 5.3 One-dimensional representation of rotation in a plane.

Angular displacement, velocity and acceleration

Rotational displacement is defined by the angle turned through. Displacement of any point is given by this angle multiplied by the radial distance of the displaced point from the axis of rotation. This is

$$x = r\theta$$

where x is the displacement along a circular path, r is the radial distance of the point from the axis and θ is the angle turned through. The displacement follows a circular path and (unlike displacement in a straight line) it is repetitive, going continuously around the same path. Because an angle has no dimensions the displacement has units of length the same as the radius.

Angular speed is rate of change of angle and angular velocity is angular speed in a specified direction. A point following a circular path at radial distance r from the axis of rotation has a velocity along this path of

$$S = r\omega$$

where ω is the angular velocity. The linear velocity of the point is continuously changing because its direction is continuously changing. It has two components: $v_T = r\omega$ in a direction tangential to the radius, and v_R along the radius towards the axis of rotation.

Angular acceleration is rate of change of angular velocity. The acceleration of a point moving along a circular path is

$$a = r\dot{\omega}$$

where $\dot{\omega}$ is the angular acceleration. As with velocity, linear acceleration of the point is continuously changing and has two components: $a_T = r\dot{\omega}$ tangential to the radius, and a_R along the radius towards the axis of rotation.

Moment, torque and inertia

A moment is defined as force multiplied by distance, where the force is at right angles to the distance as shown in Fig. 5.4(a). This is known as a cross product, and it differs from the work done when a force moves in a straight line. The latter is known as a scalar product and is shown in Fig. 5.4(b), with force and displacement in the same direction. The units in each case are newton metres but, to reduce the chance of confusion, the unit of work is given the distinctive name of the joule (J).

A moment can be present without any displacement occurring and so with

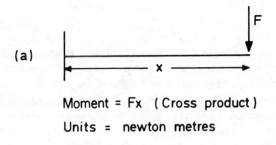

Moment = Fx (Cross product)
Units = newton metres

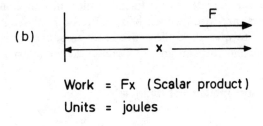

Work = Fx (Scalar product)
Units = joules

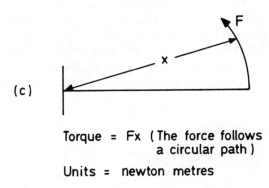

Torque = Fx (The force follows
a circular path)
Units = newton metres

Figure 5.4 (a) Moment, (b) work and (c) torque.

no work done. If the moment causes displacement work is done and is given by the product of the force and the distance through which it moves. If the force remains perpendicular to the radius it follows a circular path and the work done is $Fr\theta$, where r is the radius and θ the angle the radius turns through. Such a force is known as a torque and is shown in Fig. 5.4(c).

If the force remains in its original direction, the angle between it and the radius changes. The moment is $F\sin\theta$ and the work done is $Fr\sin\theta$. There is also a component of force $F\cos\theta$ along the radius, although if the radius remains constant this does no work.

For a rigid body rotating about one axis, Newton's second law of motion can be expressed as

$$T = I\dot{\omega}$$

where T is torque, I is moment of inertia and $\dot{\omega}$ is angular acceleration.

The moment of inertia, which is the equivalent of mass in linear motion, is a function of the shape of the body as well as of its mass. It can be regarded as being produced by the mass concentrated at a fixed radius from the axis of rotation. This notional radius is called the *radius of gyration* and is usually represented by the letter k.

Comparison with linear motion

Equations of motion can be obtained from force, energy or impulse–momentum balances exactly as for linear motion in one dimension. If the moments involved are torques then behaviour is directly equivalent. If the moments are not torques then allowance needs to be made for the effects of changes in angle, and this can greatly increase the complexity.

(b) Combined linear motion and rotation

Equivalent mass

Combined linear motion and rotation is resolved into linear motion of the centre of gravity of a rigid body plus rotation about the centre of gravity. A simple case to consider is an unrestrained uniform rod subjected to an impulse Ft perpendicular to its longitudinal axis at a distance l from the centre of gravity. If the resulting linear velocity of the centre of gravity is v_{CG} and the resulting angular velocity is ω, then the linear velocity at the point of impact is

$$v = v_{CG} + \omega l$$

Equating impulse to momentum,

$$Ft = Mv_{CG}$$

and

$$Ftl = I\omega$$

where I is the moment of inertia of the rod about its centre of gravity. Combining these gives

$$Mv_{CG} = \frac{I\omega}{l}$$

$$= \frac{Mk^2\omega}{l}$$

where k is the *radius of gyration* about the centre of gravity. Then

$$v_{CG} = \frac{k^2\omega}{l}$$

and

$$v = \omega\left(\frac{k^2}{l} + l\right) \tag{5.37}$$

In calculations it is often convenient to regard the rod as a simple equivalent mass M_E at the point of application of the force. By definition,

$$M_E v = M v_{CG}$$

giving

$$M_E = \frac{Mv_{CG}}{v} = M\frac{v_{CG}}{v_{CG} + \omega l} = M\frac{k^2}{k^2 + l^2} \tag{5.38}$$

It only applies to small rotations of the rod.

Centre of percussion
There will be a point in the rod which is the instantaneous centre of rotation during the application of the impulse, and this will remain stationary. This point is called the *centre of percussion*, and if the rod is pivoted about this point there is no reacting force at the pivot. If the centre of percussion is located a distance λ from the centre of gravity, and impulse is equated to momentum,

$$Ftl = I\omega \quad \text{for rotation about the centre of gravity}$$
$$Ft(l + \lambda) = I_{CP}\omega \quad \text{for rotation about the centre of percussion}$$

where I_{CP} is the moment of inertia of the rod about the centre of percussion. By standard theory,

$$I_{CP} = I + M\lambda^2$$

Then, by eliminating Ft,

$$\frac{I\omega}{l} = \frac{(I + M\lambda^2)\omega}{l + \lambda}$$

Therefore, using the radius of gyration,

$$\frac{Mk^2}{l} = \frac{M(k^2 + \lambda^2)}{l + \lambda} \tag{5.39}$$

This gives $\lambda = 0$ or $\lambda = k^2/l$.

Non-axial impact of a rigid rod

Analysis of combined linear motion and rotation can become complicated because of the number of variables involved. As a simple example the non-axial impact of a rigid rod with a flat rigid anvil is discussed briefly. This is shown in Fig. 5.5. It is assumed that the anvil is frictionless so that the only force present is perpendicular to the anvil surface. The mass of the rod is M, its moment of inertia about its centre of gravity is $I = Mk^2$, and the centre of gravity is distance l from the impact end of the rod.

At any time after initial contact there are three unknowns: the velocity v of the rod's centre of gravity perpendicular to the anvil; the angular velocity ω of the rod; and the angle θ between the rod and the anvil. Three equations can be derived – impulse–momentum balances for linear motion and rotation, and a geometrical relationship among v, ω and θ. These are

$$M(u - v) = Ft \tag{5.40}$$

where u is the initial velocity of the rod perpendicular to the anvil,

$$I\omega = Ftl\cos\theta \tag{5.41}$$

and

$$v = \omega l \cos\theta \tag{5.42}$$

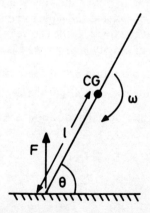

Figure 5.5 Non-axial impact of rod with frictionless anvil.

From Equations 5.40 and 5.41,

$$Ml\cos\theta(u - v) = Mk^2\omega$$

Substituting from Equation 5.42,

$$Ml\cos\theta(u - v) = \frac{Mk^2v}{l\cos\theta}$$

giving

$$v = \frac{ul^2\cos^2\theta}{k^2 + l^2\cos^2\theta} \tag{5.43}$$

and hence ω by substitution in Equation 5.42.

If the anvil is not frictionless and the coefficient of friction between the rod and the anvil is μ, then Equation 5.41 becomes

$$I\omega = Ftl(\cos\theta - \mu\sin\theta)$$

but otherwise the analysis remains the same.

5.3 MORE DETAILED ANALYSIS

More detailed analysis consists basically of extension of the analysis methods outlined above to two or three dimensions. There is a considerable body of published work on the dynamics of rigid bodies in three dimensions which goes into a large number of cases in considerable detail, although only a small proportion of these deal with impact. These analyses involve the manipulation of tensors, vectors and matrices and are beyond the scope of this book. Condensed notation and analysis are used which are basically the same as those outlined for three-dimensional stress analysis in Chapter 1. There are a large number of variables involved in three-dimensional analysis – for instance, a simple impulse–momentum balance for two rigid bodies colliding leads to twelve fairly complicated equations – and it is usually necessary to use a computer.

6
Undamped linear vibrations

6.1 INTRODUCTION

6.1.1 General

(a) Scope of chapter

The vibration of a single-mass undamped linear system has been widely studied because it demonstrates, in its simplest form, the general behaviour of vibrating bodies. When it is applied to impact, simple linear vibration analysis is often confined to the first quarter-cycle of the response, by the end of which the vibrating body has reached a maximum displacement. In such cases the analysis can be applied equally well to bodies which deform plastically and do not vibrate, as long as resistance increases linearly with displacement as it would with a linear elastic spring.

The most widely known type of vibration is steady-state behaviour under a sinusoidally varying excitation of constant peak amplitude and constant frequency. Transient vibrations are less widely known but have also been studied extensively. Behaviour under impact is discussed in the context of steady-state and transient vibration theory.

(b) Standard pulse shapes

The shape of the force–time pulse applied to a vibrating system can have a marked effect on its response. Actual pulses may be very complex and analysis is often carried out with simple standard pulse shapes. These are chosen partly to simplify the analysis and partly to highlight or minimize certain aspects of behaviour. As a result, different standard pulse shapes are used in different types of analysis.

Standard pulse shapes often either are symmetrical or incorporate a sharp step at one end of the pulse. Typical pulse shapes and the areas in which they are used are given in Table 6.1.

Table 6.1 Standard simple force–time pulses.

Pulse shape	Application
Triangle with instantaneous rise.	Explosive and blast loading. Often used with double asymptotic presentation.
Triangle with instantaneous decay.	Tests for residual response of vibrating systems. Produces more consistent results than most other pulse shapes.
Symmetrical triangle or half sine.	Simple symmetrical pulses used when pulse shape is unknown or is approximately symmetrical.

6.1.2 Basic relationships

(a) Duration of impact

Any deformable system has a characteristic response time. With a vibrating system it is the natural period of the vibration. The response to a force–time pulse depends on the duration of the pulse relative to the characteristic response time. This behaviour varies in detail but, as a rough guide, if the pulse lasts for about one-tenth of the response time or less, it can usually be treated as an impulse producing an instantaneous velocity change. If the pulse reaches its maximum value in about one-tenth of the response time or less, and then remains at this value for some time, the pulse can usually be regarded as a step input producing an instantaneous acceleration.

Otherwise it is necessary to consider the shape of the pulse. Four or more different times may be needed in the analysis and there is ample scope for confusion. These include the response time T of the body, the duration of the pulse t_d, the time τ at which some specific force amplitude was reached and the time t at which behaviour is being considered. In some cases it may be convenient to introduce additional times such as t_{max}, the time at which maximum displacement occurs.

(b) Initial and residual response

There are two main types of response to a pulse load which have been studied extensively: the maximum response, which often occurs at the end of the first quarter-cycle of vibration; and the residual vibration remaining after the pulse is over. The first is often given in the form of an *initial shock spectrum* and the second in the form of a *residual shock spectrum*. The latter can be very sensitive to small changes in the duration of the impact, and in tests it is necessary to use pulse shapes which minimize this sensitivity if consistent results are to be obtained.

6.2 LINEAR MOTION IN ONE DIMENSION

6.2.1 Vibrations

(a) Free vibrations

The model representing a single-mass linear undamped vibrating system consists of a mass M attached to one end of a massless linear spring of *spring stiffness* K, the other end of which is fixed rigidly. When the system is at rest, no forces act on it and no strain energy is stored in the spring. If the mass is displaced and released or is given an initial velocity, it vibrates. The force exerted on the mass by the spring is $F = \pm Kx$ and the inertia force exerted on the spring by the mass is $-F = \pm Ma$, where x is the displacement and a is the acceleration. Eliminating F gives

$$Ma + Kx = 0 \tag{6.1}$$

This is a standard differential equation. If $x = 0$ at $t = 0$,

$$x = A \sin \omega t \tag{6.2}$$

where $\omega = \sqrt{(K/M)}$ and A is an arbitrary constant. If $x = A$ at $t = 0$,

$$x = A \cos \omega t$$

and the full solution for any initial displacement x_0 and initial velocity v_0 can be shown to be

$$x = \frac{v_0}{\omega} \sin \omega t + x_0 \cos \omega t \tag{6.3}$$

For simplicity further discussion is restricted to Equation 6.2, but it is equally valid for the full Equation 6.3.

Equation 6.2 represents a free vibration of the mass, which executes simple harmonic motion with a frequency $f = \omega/2\pi$ and a period $T = 2\pi/\omega$. These are called the natural frequency and natural period of the system and are independent of the value of A.

The same result can be obtained from an energy balance between the strain energy in the spring and the kinetic energy of the mass. The result can also be derived from the Lagrange equation of motion (Equation 5.21).

The velocity v and the acceleration a of the mass are obtained by differentiating displacement x with respect to time. This gives

$$v = \frac{dx}{dt} = -A\omega \cos \omega t \tag{6.4}$$

$$a = \frac{dv}{dt} = -A\omega^2 \sin \omega t \tag{6.5}$$

The relationships among the maximum values of x, v and a vary with ω as shown in Fig. 6.1. In Fig. 6.1(a) x is taken as constant, but if v is taken to be

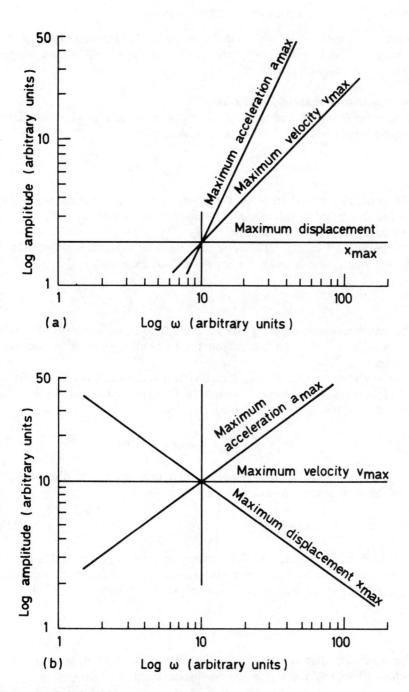

Figure 6.1 Relationships among maximum vibration displacement, velocity and acceleration as angular velocity varies. (a) Constant displacement. (b) Constant velocity.

constant as in Fig. 6.1(b) it is possible to plot a symmetrical carpet graph linking x, v, a and ω. Such graphs are widely used, but strictly they only apply to simple harmonic motion.

(b) Steady-state sinusoidal excitation

If the mass is not allowed to vibrate freely but is subjected to a sinusoidally varying force $F \sin \omega_1 t$, the displacement becomes

$$x = \frac{F/K}{1 - (\omega_1/\omega)^2} \left(\sin \omega_1 t - \frac{\omega_1}{\omega} \sin \omega t \right) \tag{6.6}$$

This contains a forced vibration component $\sin \omega_1 t$ at the frequency of the applied force, and a free vibration component $(\omega_1/\omega) \sin \omega t$ at the natural frequency of the vibrating system.

In the standard theory of steady-state forced vibrations it is assumed that the free vibration dies away, leaving the displacement as

$$x = \frac{F/K}{1 - (\omega_1/\omega)^2} \sin \omega_1 t \tag{6.7}$$

The maximum displacement A varies with the ratio ω_1/ω, as shown in Fig. 6.2. When $\omega_1 = \omega$ resonance occurs and the theoretical value of A becomes infinite.

An alternative way in which a forced vibration can occur is when the fixed end of the spring is subjected to a sinusoidally varying displacement $x_s \sin \omega_1 t$. The absolute displacement of the mass is now

$$x_{\text{abs}} = \frac{x_s}{1 - (\omega_1/\omega)^2} \left(\sin \omega_1 t - \frac{\omega_1}{\omega} \sin \omega t \right) \tag{6.8}$$

which is the same as Equation 6.6 with F/K replaced by x_s. The displacement x_r of the mass relative to the other end of the spring is

$$x_r = \frac{x_s}{(\omega/\omega_1)^2 - 1} \left(\sin \omega_1 t - \frac{\omega}{\omega_1} \sin \omega t \right) \tag{6.9}$$

In different circumstances the displacement of interest may be x, x_{abs} or x_r.

In standard steady-state forced vibration theory it is assumed that the free vibration dies away, leaving the maximum displacements as

$$A_{\text{abs}} = \frac{1}{1 - (\omega_1/\omega)^2} \qquad A_r = \frac{1}{(\omega/\omega_1)^2 - 1}$$

The maximum displacements in the positive and negative directions are numerically equal, and so the relative displacement can be rewritten

$$A_r = \frac{1}{1 - (\omega/\omega_1)^2} \tag{6.10}$$

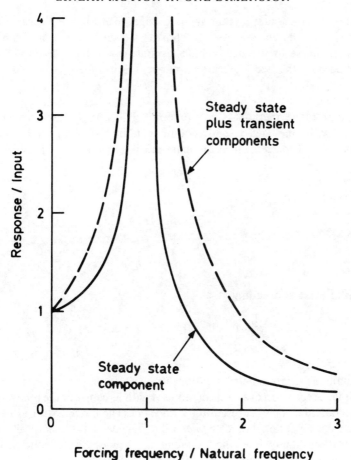

Figure 6.2 Variation in response with sinusoidal forcing frequency for systems with one degree of freedom.

This varies exactly as A does in Fig. 6.2 but with the units on the horizontal axis inverted.

(c) Short duration excitation

Sinusoidal transients
When a sinusoidally varying force with a constant peak amplitude is first applied it takes time for the vibration to build up to its steady-state value. If the force is applied for an integral number of half-cycles, A varies with both ω_1/ω and the number of half-cycles N within an envelope bounded by the curve $A = (F/K)/[1 - (\omega_1/\omega)]$. This envelope can be obtained simply from

Equation 6.6 by assuming that at some stage $\sin \omega_1 t = 1$ at the same time as $\sin \omega t = -1$. The envelope is also shown as a dashed line in Fig. 6.2. It is similar in shape to the steady-state response curve but the amplitudes are larger because the free vibration component is included.

Phase-plane plots

The phase-plane method is a geometric analysis applicable to vibrating systems with one degree of freedom, including damped and non-linear systems. The displacement of the mass in an undamped linear system is

$$x = A \sin \omega t$$

and its velocity is

$$v = A\omega \cos \omega t$$

giving

$$\frac{v}{\omega} = A \cos \omega t$$

Squaring these and adding gives

$$x^2 + \left(\frac{v}{\omega}\right)^2 = A^2(\sin^2 \omega t + \cos^2 \omega t)$$

$$= A^2$$

by a standard trigonometric relationship.

This is the equation of a circle of radius A with its centre at the intersection of two rectangular axes representing x and v/ω. The circle is traced out by a radius rotating at angular velocity $\omega = 2\pi/T$, where T is the natural period of the vibration. At any time t the end of the radius gives the co-ordinates of both x and v/ω (Jacobson and Ayre, 1958, pp. 145–51).

A force–time pulse can be approximated by a number of small step changes in force and the response can be derived graphically. The method automatically gives the end conditions for each step and these can be used directly to give the starting conditions for the next step.

Impulse–momentum analysis

Impulse–momentum analysis can be used to give the free vibration response to a force–time pulse of arbitrary shape. The convolution integral for a rigid body, given by Equation 5.34, can be extended to the simple, linear, undamped vibrating system and it can be shown that

$$x = \frac{1}{M\omega} \int_0^t F \sin \omega t(t - \tau) \, \mathrm{d}\tau \tag{6.11}$$

where F is the applied force, which can vary with time, τ is the time at which F

has any specific value and t is any time later than τ. The equation can also be expressed in terms of spring stiffness using the relationship

$$\frac{1}{M\omega} = \frac{\omega}{K} \qquad (6.12)$$

which is derived by rewriting $\omega = \sqrt{(K/M)}$.

Equation 6.3 for the free vibration response to an initial velocity and displacement can be added to Equation 6.11 to give

$$x = \frac{v_0}{\omega}\sin \omega t + x_0 \cos \omega t + \frac{\omega}{K}\int_0^t F \sin \omega(t-\tau)\,d\tau \qquad (6.13)$$

This is the comprehensive expression for the displacement of a single-mass undamped linear vibrating system using the theory of free vibrations. It is sometimes more convenient to rewrite the equation as (Biggs, 1964, pp. 40–2)

$$x = \left(x_0 - \frac{\omega}{K}\int_0^t F \sin \omega\tau\,d\tau \right)\cos \omega t$$

$$+ \left(\frac{v_0}{\omega} + \frac{\omega}{K}\int_0^t F \cos \omega\tau\,d\tau \right)\sin \omega t \qquad (6.14)$$

Fourier analysis
Fourier analysis is usually used to apply the theory of steady-state vibrations to the vibration response to a force–time pulse of arbitrary shape. As discussed in Section 6.2.1(b) there are differences between steady-state vibrations and free vibrations, and so this approach is not strictly correct. To balance this the theory is widely known and standard analysis techniques are readily available.

A *Fourier series* can be derived for a simple force–time pulse by taking the limiting case in which the time between repetitions is infinite and the fundamental frequency is zero. There are then an infinite number of *harmonics*. There is a zero frequency difference between adjacent harmonics; thus they cannot be plotted separately but must be plotted as an amplitude density. The horizontal axis represents frequency, the area under the curve represents amplitude and the vertical axis represents amplitude per unit of frequency, which is called *amplitude density*. The total amplitude in any frequency band is the area of the strip bounded by the upper and lower frequency of the band. The amplitude density plot is called the *Fourier transform* or the Fourier spectrum of the pulse.

In its complete form a Fourier transform contains enough information to reconstitute the pulse completely. It can be a very complicated expression incorporating positive and negative frequencies, positive and negative amplitude densities and real and imaginary components. In simple analysis it is usual to take only positive frequencies, positive amplitude densities and real components. This simplifies analysis considerably, but it is not possible to reconstitute the original pulse because information such as the phase relationships has been lost.

The simple transform for a rectangular amplitude–time pulse can be shown to be

$$\mathscr{F} = At_\mathrm{d} \left| \frac{\sin \pi f t}{\pi f t} \right|$$

where A is the amplitude, t_d is the pulse duration and f is the frequency of each harmonic. The parallel vertical lines indicate that the quantity between them is always given a positive value.

The transform is plotted dimensionlessly in Fig. 6.3(a). In order to apply the transform to a specific pulse, the vertical scale is multiplied by At_d and the horizontal scale by $1/t_\mathrm{d}$. The result is shown in Fig. 6.3(b) for three pulses of equal amplitude but different duration. As duration increases the transform becomes taller and narrower until, with an instantaneous step load, which can be regarded as a pulse of infinite duration, it becomes a Dirac spike of infinite height and zero width.

The response function of the vibrating system is taken to be its steady-state response to constant amplitude sinusoidal inputs. Multiplying the Fourier transform of the input by the response function gives the Fourier transform of the output. This can then be used to give the output response, but with the same loss of detail as occurred in making the transform of the input.

This loss of detail and the inaccuracies introduced by using the steady-state response to analyse transient behaviour are usually not important. The parameters of the input are seldom known accurately and theoretical results need to be related to actual results by the use of empirical correction factors (Broch *et al.*, 1980, Chapter 2; Barber, 1961).

6.2.2 Simple force–time pulses

(a) Constant force inputs

With a constant force input the applied force reaches a maximum value and then remains constant. Two simple cases are discussed: a step input, in which the force is applied instantaneously; and a ramp input, in which the force increases linearly with time up to its maximum value.

With the step input the force balance equation is

$$Ma + Kx = F \tag{6.15}$$

If $x = 0$ at $t = 0$ this has the solution

$$x = \frac{F}{K}(1 - \cos \omega t) \tag{6.16}$$

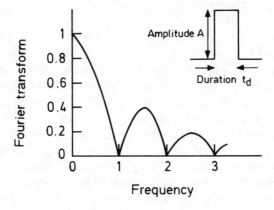

(a)

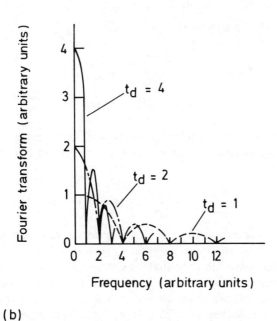

(b)

Figure 6.3 Fourier transforms of rectangular pulses. (a) Dimensionless plot applicable to all rectangular pulses. (b) Plots defining specific pulses of equal amplitude but different duration.

in terms of the spring stiffness, or

$$x = \frac{FM}{\omega^2}(1 - \cos \omega t) \tag{6.17}$$

in terms of the mass. This represents a free vibration with a peak displacement $A = \pm F/K$ about a steady mean displacement of F/K, which is the amount that the spring would deflect under a static force F. In the direction of the applied force the maximum deflection is twice the deflection for the same load applied statically. In the opposite direction the deflection is zero. Note that the steady mean displacement has no effect on velocity and acceleration.

If the force increases linearly with time, reaching its maximum value F_{max} at time t_d, there are two stages to the analysis. In the first stage $x_0 = 0$ and $v_0 = 0$. The differential equation of motion becomes

$$M_a + Kx = \frac{F_{max}}{t_d}t \tag{6.18}$$

and Equation 6.13 becomes

$$x = \frac{\omega}{K}\int_0^t \frac{F_{max}}{t_d}\tau \sin \omega(t - \tau)\,d\tau \tag{6.19}$$

Both can be shown to give

$$x = \frac{F_{max}}{t_d K\omega}(\omega t - \sin \omega t) \tag{6.20}$$

$$v = \frac{F_{max}}{t_d K}(1 + \cos \omega t) \tag{6.21}$$

In the second stage the initial velocity and displacement are given by Equations 6.20 and 6.21 with $t = t_d$ and can be used in Equation 6.3. From this,

$$x = \frac{F_{max}}{K}\left\{1 + \frac{1}{t_d\omega}[\sin \omega(t - t_d) - \sin \omega t]\right\} \tag{6.22}$$

The value of x varies with $1/t_d\omega$. It may be more convenient to use the natural period of the vibration $T = 2\pi/\omega$ so that x varies with $T/2\pi t_d$. As t_d/T decreases the response becomes closer to that for a step force, and as t_d/T increases the response becomes closer to that for the force F_{max} applied statically. Some typical intermediate responses are shown in Fig. 6.4. The change in response does not occur uniformly with change in t_d/T but depends also on how close t_d is to being a multiple of T. The maximum displacement is shown in Fig. 6.5. If t_d is less than $T/4$ the maximum displacement is within 10% of the value for a step force, and if t_d is more than $3T$ the maximum displacement is within 10% of the static value (Jacobson and Ayre, 1958, Chapter 4).

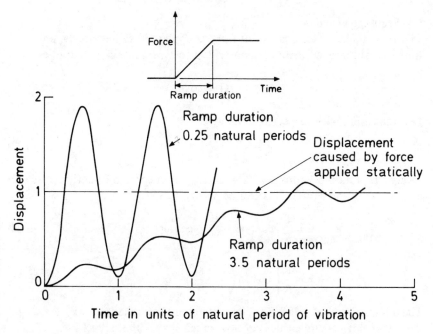

Figure 6.4 Response to ramp-shaped application of a constant force.

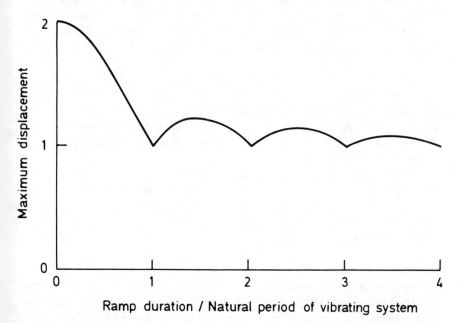

Figure 6.5 Variation in maximum displacement with ramp duration.

(b) Rectangular pulses

In a rectangular pulse a constant force is applied instantaneously, and after time t_d it is removed instantaneously. Up to t_d, Equation 6.16 applies:

$$x = \frac{F}{K}(1 - \cos \omega t)$$

Differentiating with respect to time gives

$$v = \frac{F\omega}{K} \sin \omega t \qquad (6.23)$$

After t_d, Equation 6.3 applies with x_0 and v_0 obtained from Equations 6.16 and 6.23 at $t = t_d$ and $F = 0$. This can be shown to give

$$x = \frac{F}{K}[\cos c \cdot (t - t_d) - \cos \omega t] \qquad (6.24)$$

or

$$x = \frac{F}{K}\left[\cos 2\pi \left(\frac{t}{T} - \frac{t_d}{T} \right) - \cos 2\pi \frac{t}{T} \right] \qquad (6.25)$$

The full maximum displacement for a step load is reached when $t_d = T/2$ and then does not vary regardless of how much longer the pulse lasts.

There is usually a residual vibration remaining after the pulse ends, and this varies cyclically with the value of t_d/T. When t_d is a multiple of T the situation is as shown in Fig. 6.6(a). The force is removed when there is zero displacement and zero velocity and there is no residual vibration. This can also be derived from Equation 6.25 because $\cos 2\pi[(t/T) - n] = \cos 2\pi(t/T)$ when n is a positive integer.

If the pulse ends exactly out of phase with this, the situation is as shown in Fig. 6.6(b). The force is removed when there is zero velocity but maximum displacement and there is a residual vibration about the zero deflection position with an amplitude of $\pm x_{max}$. The maximum amplitude in the direction of the force remains as it was but there is now an equal displacement in the other direction too.

For any other value of t_d/T there is a residual vibration about the zero displacement position with an amplitude somewhere between zero and $\pm x_{max}$, as shown in Fig. 6.6(c). The way in which the maximum displacement of the residual vibration varies with t_d/T is shown in Fig. 6.7 (Kornhauser, 1967, Section 5.3).

(c) Triangular pulses

A representative triangular force–time pulse is shown in Fig. 6.8. Force increases linearly with time up to a maximum value F_{max} at time t_1 and then decreases linearly with time, reaching zero at time t_d. Response to the pulse is

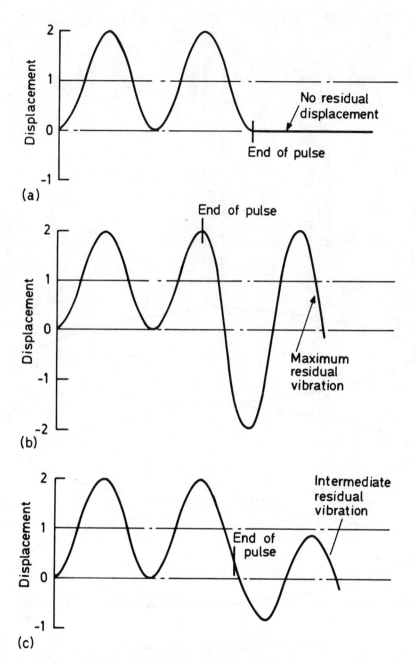

Figure 6.6 Residual vibration after a rectangular force–time pulse. (a) t_d multiple of T. (b) t_d multiple of $T/2$. (c) Other value.

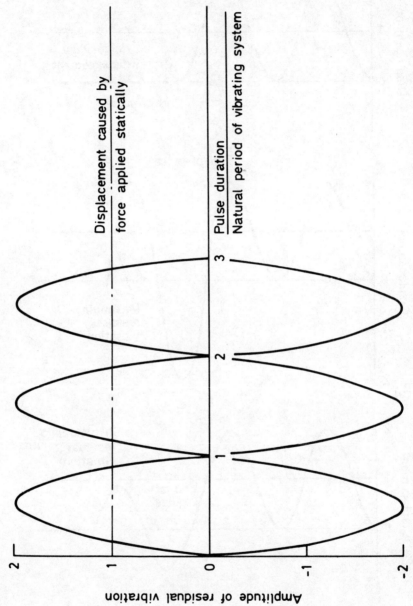

Figure 6.7 Variation in residual vibration with pulse duration for a rectangular force–time pulse. Usually only the positive half-amplitude is shown.

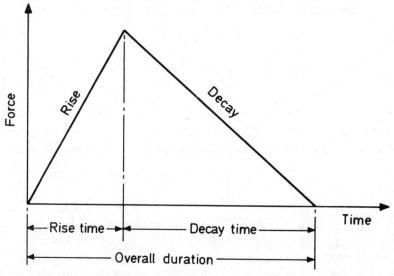

Figure 6.8 Triangular force–time pulse.

analysed in three stages. The force is given by

$$F = \frac{F_{max}t}{t_1} \qquad \text{in the first}$$

$$F = F_{max}\left(1 - \frac{t - t_1}{t_d - t_1}\right) \qquad \text{in the second}$$

$$F = 0 \qquad \text{in the third}$$

The final displacement and velocity in each stage are taken as the initial displacement and velocity for the next stage.

An important feature of a triangular pulse is its skewness, which can be represented by a *skewness factor* $S = t_1/t_d$. Apart from symmetrical pulses in which $S = 0.5$, all other triangular pulses occur in mirror image pairs; for example, the pulse with $S = 0.2$ is the mirror image of the pulse with $S = 0.8$. Triangular pulses are useful for studying the effects of skewness because both the maximum force F_{max} and the total impulse $1/2\ (F_{max}t_d)$ can be kept constant as the pulse shape changes.

For illustration, the behaviour of three pulses of the same maximum amplitude and pulse duration are discussed – a symmetrical pulse and the mirror image pair with $S = 0$ and $S = 1$. These are shown in Fig. 6.9. The pulse with $S = 0$ starts with an instantaneous application of force and approximates to the shape of pulse found with explosive and blast loading. The pulse with

Skewness factor   =   Rise time / Overall duration

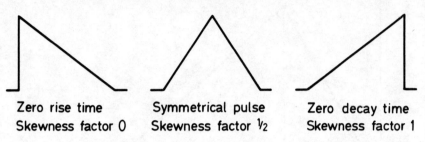

| Zero rise time | Symmetrical pulse | Zero decay time |
| Skewness factor 0 | Skewness factor ½ | Skewness factor 1 |

Figure 6.9 Skewness of triangular pulses.

$S = 1$ ends with an instantaneous removal of force and approximates to a shape of pulse widely used in the shock testing of vibrating systems.

The displacement of a single-mass linear undamped vibrating system is shown in Fig. 6.10. In each case displacement occurs more quickly with the faster rising forces. When $t_d = T/4$, as in Fig. 6.10(a), the three responses are very similar apart from a slight phase shift. However, when $t_d = 1.5T$, for example, as shown in Fig. 6.10(b), behaviour during the pulse is very different. A rather surprising result is that the residual vibration remaining after the pulse has ended is the same, apart from phase shift, for the two mirror image pulses. This is a general property of mirror image pulses not just of triangular ones (Jacobson and Ayre, 1958).

(d) Sinusoidal pulses

Half-sine
Symmetrical force–time pulses are often taken to be sinusoidal in shape. The response to a half-sine pulse is found by using the transient vibration theory discussed in Section 6.2.1(c) with only the first half-sine wave considered. As with rectangular pulses, the amplitude of the residual vibration can vary with the ratio t_d/T. It lies between zero and $1.73 F_{max}/K$.

Versed sine
The versed sine pulse is actually a cosine curve with the equation

$$F = \frac{F_{max}}{2}\left(1 - \cos\frac{2\pi t}{t_d}\right)$$

for a symmetrical pulse. It covers a full cycle of a sinusoidal wave. Asymmetrical pulses are also used in which the duration of the first half-cycle differs from that of the second half-cycle.

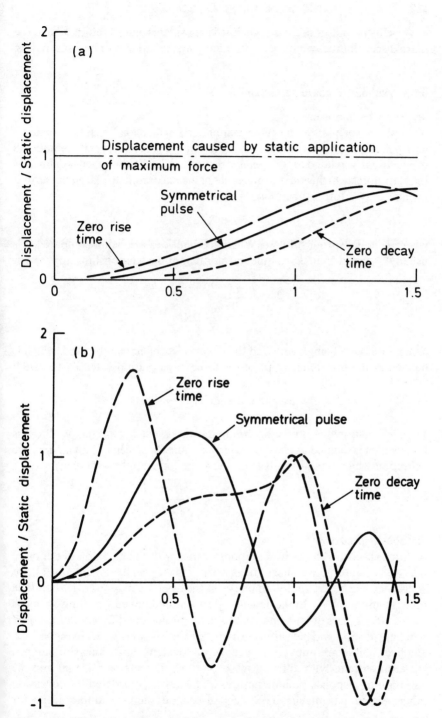

Figure 6.10 Effect of skewness and duration of triangular pulse. (a) Overall pulse duration 0.25 natural periods. (b) Overall pulse duration 1.5 natural periods.

Versed sine pulses are used when it is considered undesirable to have the sharp discontinuities which occur in triangular, rectangular or half-sine pulses.

6.2.3 Methods of characterization

(a) Static comparisons

A simple way of quantifying behaviour under dynamic loading is to compare it with behaviour under a reference static loading. There are several methods, all of which can be related to the *dynamic response index (DRI)*, which is defined as the static force which would produce the same maximum displacement as that produced by a force–time pulse. This gives

$$DRI = Kx_{max}$$

where K is the spring stiffness. When accelerations are being compared it is usually assumed that the simple harmonic motion relationships apply with

$$a_{max} = \omega^2 x_{max}$$

$$= \frac{\omega^2}{K} DRI$$

When forces are being compared the *dynamic magnifier* can be used, which is defined as the maximum value of the force–time pulse divided by the DRI:

$$\text{dynamic magnifier} = \frac{F_{max}}{Kx_{max}}$$

When displacements are being compared the *dynamic load factor (DLF)* is used, which is defined as the dynamic maximum displacement divided by the deflection which would result from the same force applied statically:

$$DLF = \frac{x_{max}}{F_{max}/K} = \frac{DRI}{F_{max}}$$

(b) Shock spectra

A shock spectrum shows how response varies with the duration of the pulse relative to the response time of the system. This is usually done in one of two ways; these are comparable, but one gives the response in terms of t_d/T and the other in terms of T/t_d. In the first a single linear undamped vibrating system is subjected to pulses of the same shape and amplitude but different durations; T remains constant and the variation in pulse length is considered. In the second, identical pulses are applied to a range of constant mass, linear undamped vibrating systems with different response times; t_d remains constant and the variation in response time or natural frequency is considered. On occasion other related parameters have been used and there is ample scope for

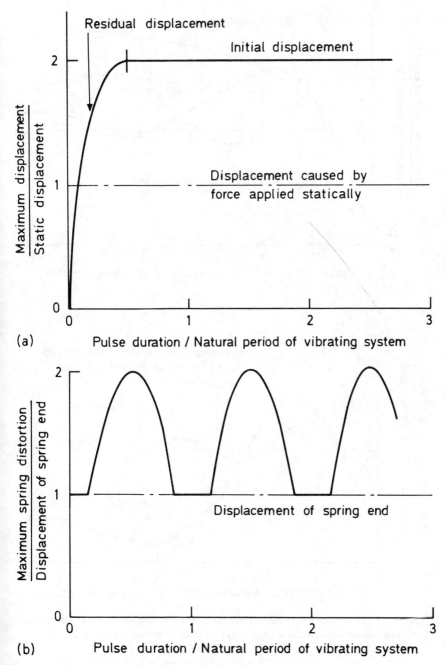

Figure 6.11 (a) Maximax displacement spectrum for rectangular force–time pulse.
(b) Maximax spring distortion for rectangular displacement–time pulse.

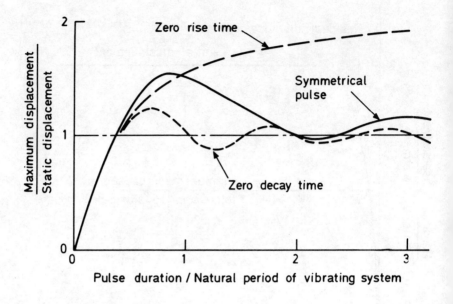

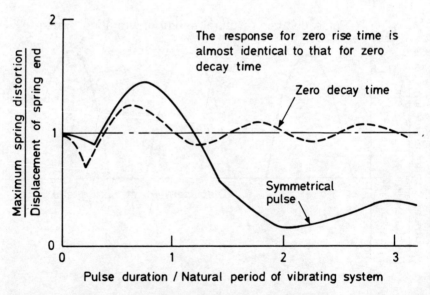

Figure 6.12 (a) Maximum displacement spectrum for triangular force–time pulse. (b) Maximum spring distortion for triangular force–displacement pulse.

confusion. It is essential to check the relative pulse duration parameter before using published shock spectra. Once this scale is understood, all shock spectra give equivalent results.

Usually maximum displacement is used to represent the response. There are three ways of defining the maximum displacement. The *initial shock spectrum* uses the displacement at the first peak which often occurs at the end of the first quarter-cycle of vibration. The *residual shock spectrum* uses the displacement during subsequent vibration, and the *maximax shock spectrum* uses the overall maximum displacement. In addition there are two ways in which displacement can be measured. The *displacement spectrum* uses the maximum displacement of the mass, and the *distortion spectrum* uses the maximum deflection of the spring. These are the same if a force pulse is applied to the mass but different if a displacement pulse is applied to the fixed end of the spring.

For illustration, the response to a rectangular force–time pulse is discussed. The maximax displacement spectrum is shown in Fig. 6.11(a). This is the same as the residual spectrum for pulses shorter than $0.5T$ and as the initial spectrum for pulses longer than this. The residual spectrum is the same as the simple Fourier transform for the rectangular pulse with the amplitudes multiplied by $2\pi f$. It can be shown that this applies for any force–time pulse. For comparison the maximax spring distortion is shown in Fig. 6.11(b) for a rectangular displacement–time pulse applied to the fixed end of the spring. It is similar to the residual displacement spectrum for a force–time pulse applied to the mass with the lower values omitted.

Maximax displacement and distortion spectra for the triangular pulses discussed in Section 6.2.2(c) are shown in Fig. 6.12(a) and (b) (Jacobson and Ayre, 1958).

(c) Limiting envelopes

Shock spectra can be used for the maximum value of displacement, velocity or acceleration, and the one which is important can vary with the frequency being considered. At low frequencies accelerations are low and the limiting factor is usually the maximum displacement. At high frequencies the displacements are low and the limiting factor is usually the maximum acceleration. At intermediate frequencies the maximum velocity may be important.

All three parameters can be plotted together. This is usually done on a logarithmic carpet graph of the type shown in Fig. 6.1(b). Usually simple straight line plots are used, as shown in Fig. 6.13. The limiting envelope is taken as the lowest spectrum at any frequency. Often the circular frequency ω is used, giving the relative duration scale as $t_d/T = t_d\omega/2\pi$.

(d) Double-asymptote comparisons

A drawback of comparisons based on relative pulse duration or relative frequency is that, although the reference scale goes to infinity in one direction,

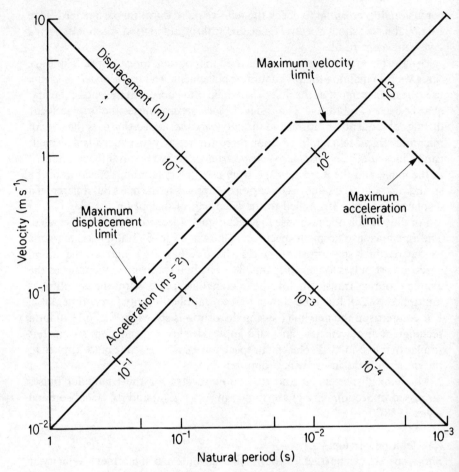

Figure 6.13 Envelope of limiting conditions.

it goes to zero in the other. Variations in response can be seen clearly for pulses in which the response curve asymptotically approaches a limiting value, but as the reference scale approaches zero the response curve becomes more and more compressed. This has led to the use of double-asymptote comparisons in which the reference scales go to infinity in both directions.

In its basic form this is done by plotting the maximum displacement against both total velocity change Δv and average acceleration a_{av}. The relationship between these is $\Delta v = a_{av}t$, and plotting this on axes of Δv and a_{av} gives a fan of straight constant time lines each representing a constant ratio of $\Delta v/a_{av}$. On log-log scales the constant time lines are parallel and at 45° to the other two axes. The graph can be made dimensionless by writing

$$v = \frac{\Delta v}{v_0}, \qquad a = \frac{a_0 T}{v_0}, \qquad t = \frac{t_d}{T}$$

where v_0 and a_0 are reference values of velocity and acceleration, which are taken to be the values for simple harmonic motion:

$$v_0 = \omega x_{max} \qquad a_0 = \omega^2 x_{max}$$

The relationship between a_0 and a_{av} depends on two factors: the relationship between average force and peak force, which is a function of pulse shape; and the relationship between peak force and x_{max}, which is given by graphs of the types shown in Figs 6.4 and 6.5. Combining these gives

$$a_{av} = \frac{a_0 \times \text{shape factor}}{\text{dynamic load factor}}$$

$$= a_0 \times \text{shape factor} \times \text{dynamic magnifier}$$

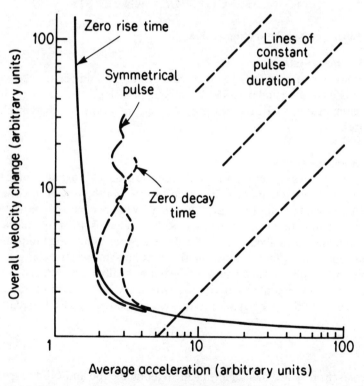

Figure 6.14 Triangular force–time pulses of different amplitudes, skewnesses and durations to give a constant maximum displacement.

Results for the three triangular pulses discussed in Section 6.2.2(c) are plotted in this way in Fig. 6.14. For short duration pulses the responses all approach a common asymptote in which behaviour is substantially unaffected by large variations in average acceleration. For long duration pulses the three curves are different but each approaches an asymptote in which behaviour is substantially unaffected by large changes in overall velocity change.

The curve for the pulse with an instantaneous application of load is particularly simple and approximates to a hyperbola. This has been found to occur for all shapes of pulse in which the maximum force occurs instantaneously at the beginning of the pulse. Such pulses occur in explosive and blast loading, and the double-asymptote method has been found useful for those. Often it is convenient to multiply both the velocity change and average acceleration axes by the vibrating mass M and to regard them as impulse and average force axes. (See Kornhauser, 1967, Section 5.4; Abrahamson and Lindberg, 1972).

6.3 MORE COMPLEX ANALYSIS

6.3.1 Linear motion in one dimension

(a) Multimass systems

The behaviour of a vibrating system consisting of a number of masses and springs is basically similar to the behaviour of a single-mass system, but is more complicated and becomes progressively so as the number of masses increases.

(b) Continuous systems

In most mechanical systems the mass and elastic properties are distributed and not lumped in discrete parts as has been assumed in the rest of this chapter. Often it is possible to approximate the system by lumped masses connected by springs, but sometimes the system has to be treated as a continuous whole. This can make the analysis more difficult because, although the lumped mass systems involve ordinary differential equations, the distributed systems involve partial differential equations which are usually more difficult to solve.

Analysis of the longitudinal vibrations of a rod of uniform cross-section involves use of the wave equation discussed in Section 1.3.2. The equation is

$$\frac{\partial^2 u}{\partial t^2} = C^2 \frac{\partial^2 u}{\partial x^2}$$

where u is the displacement of any element along the rod and C is the speed of elastic wave propagation. The appropriate solution for the free vibration of a rod unrestrained at each end is

$$u = \sum_{0}^{n} \cos \frac{n\pi x}{l} (A_n \cos \omega_n t + B_n \sin \omega_n t)$$

where ω_n is the natural frequency of a vibration mode, l is the rod length, A and B are constants and $n = 1, 2, 3, \ldots$.

The response to various types of applied for various end conditions of the rod can be derived in a way analogous to the simple single-mass vibrating system, but the mathematics involved are beyond the scope of this book (Jacobson and Ayre, 1958, pp. 466–7).

6.3.2 Three dimensions

The methods of analysis discussed above can be extended to cover linear motion in three dimensions, angular motion in three dimensions and combined linear and angular motion. The results follow the same general behaviour as in the one-dimensional case but there are so many variables that analysis is confined to specific cases. The calculations involved can fairly easily grow to an extent where computing costs and program reliability are the limitations on what it is feasible to do.

7

Deformable bodies

7.1 INTRODUCTION

7.1.1 General

(a) Scope of chapter

Bodies can deform on impact in a number of ways. More than one of these can be present at the same time, the pattern of deformation may change during the impact, and different methods of analysis may need to be combined. In this chapter different types of impact response are considered separately.

If stress waves can be ignored, published work concentrates on two main types of behaviour – the deformation of a body loaded by a given force–time pulse, and the effect of a rigid body impacting a deformable one. Only the first of these is considered here; the second is discussed in Chapter 8. Analysis is basically an extension of the methods discussed in Chapters 5 and 6.

If the duration of impact is less than, or close to, the time taken by a stress wave to traverse the body, stress waves are likely to be of interest. Usually they are treated separately from the deformation of the body as a whole. The analysis given here is basically an extension of that discussed in Chapter 1.

Most deformations occurring during an impact are irreversible. The effects of energy dissipation and damping are discussed in the context of damped vibrations. Analysis including damping is appreciably more complicated, and it is often more convenient to use an elastic analysis and to make subsequent empirical corrections for damping.

(b) Repeatability and accuracy

In theoretical analysis results are consistently repeatable. Many real-life impacts involve complicated chains of events and can only be regarded as repeatable in a statistical sense, i.e. with broad overall agreement but considerable variation in detail. It may then be better to apply a simple analysis to a representative range of impacts than to analyse a single impact in more detail. Theoretical analysis can give considerable insights into patterns of behaviour but can also give a false sense of repeatable behaviour where this does not exist.

In many impacts analysis has to be based on the permanent deformations remaining after the impact. As the discussion in Section 5.2.1 shows, very similar final deformations can result from substantially different force–time pulses. Thus there are severe limitations on the accuracy with which assumed behaviour during the impact can be verified if only the final deformations can be measured.

7.1.2 Basic relationships

(a) Energy loss and damping

Most of the deformations occurring on impact are non-conservative to a greater or lesser degree. The resulting loss of energy can be treated in various ways and it is often difficult to decide which is the most realistic. As a result the method of energy dissipation is often chosen for ease of analysis rather than for realism. The simplest relationship which includes damping involves *viscous damping*, in which the damping force is directly proportional to velocity. Most published analysis has concentrated on this even though it is known not to be realistic in many cases. *Hysteretic damping*, where the damping force can vary with displacement but is independent of velocity, and *Coulomb damping*, in which the damping force is constant, have also been studied but to a lesser extent.

The simplest models of systems which deform non-conservatively on impact consist of a mass attached to a deflecting element which contains a linear spring and a viscous damper either in series or in parallel, as shown in Fig. 7.1. Depending on the relative values of the spring and damper forces, behaviour

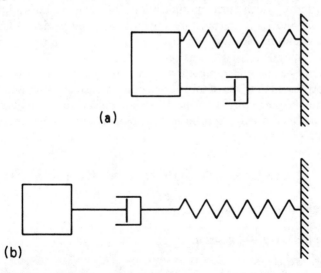

Figure 7.1 Simple models of non-conservative systems. (a) Spring and damper in parallel. (b) Spring and damper in series.

can range from completely conservative to completely unconservative. The model comes closer to the conservative undamped model as the damping force decreases with the parallel models and as it increases with the series models. Empirical relationships incorporating an assumed damping term can also be used.

Damping can increase greatly the complexity of analysis and it is often sufficient to assume a completely elastic response but to remove an appropriate amount of the stored strain energy at appropriate stages in the analysis. This can give a much simpler analysis, especially in the study of empirical results when the damping mechanism may not be known.

(b) Complexity of analysis

Comprehensive analysis of the behaviour of a deformable body on impact can become extremely complicated. Often a simple analysis will give adequate results for the complete body over a limited range of impacts or of parts of the body over a wider range. It may be simpler to plot a number of simple analyses to a common format and to take their envelope than to attempt a comprehensive analysis.

A number of methods are in common use for simplifying a complicated analysis. One is the decoupling of different aspects, as in the vibration of multimass systems where the different modes of vibration are treated separately. Another is the use of transforms, in which the variables involved are changed into more tractable variables and then the answers are transformed back into the original variables. Well-known cases of transforms are logarithms, which turn multiplication and division into addition and subtraction, and Laplace transforms, which turn differential equations into algebraic equations. Results can also be plotted in the form of charts, and numerical analysis is used extensively.

It may not be possible to present details of analysis simply, when published work is expressed in terms of transforms or numerical computations. In some parts of this chapter, although typical results are given, the methods of analysis are only outlined briefly. Appropriate references are given to more complete treatments.

7.2 LINEAR MOTION IN ONE DIMENSION

7.2.1 Single-mass systems

(a) Some basic types of resistance

Constant resistance
Where the deforming body exerts a force P the differential equation of motion is

$$Ma + P = F \tag{7.1}$$

where F is the applied force. For illustration, a constant resisting force P and a rectangular force–time pulse with amplitude F and duration t_d are considered. Whilst the force is acting,

$$a = \frac{F - P}{M} \tag{7.2}$$

giving

$$v = \frac{F - P}{M} t \tag{7.3}$$

and

$$x = \frac{F - P}{2M} t^2 \tag{7.4}$$

When the pulse has ended, but before the mass has come to rest,

$$a = -\frac{P}{M} \tag{7.5}$$

giving

$$v = \frac{F - P}{M} t_d - \frac{P}{M}(t - t_d)$$

$$= \frac{1}{M}(F t_d - P t) \tag{7.6}$$

and

$$x = \frac{F - P}{2M} t_d^2 + \frac{1}{M}(F t_d - P t)(t - t_d) - \frac{P}{2M}(t - t_d)^2 \tag{7.7}$$

Motion stops when the impulse applied to the mass has all been dissipated by the retarding force,

$$F t_d = P t_{max}$$

giving

$$t_{max} = \frac{F t_d}{P} \tag{7.8}$$

and

$$x_{max} = \frac{P t_d^2}{2}\left(1 - \frac{P}{F}\right) \tag{7.9}$$

The system does not have a characteristic response time in the way that a vibrating system does, but it may be useful to specify a characteristic time of some sort. One such time is the total duration of motion T when the mass is subjected to a pure impulse of finite magnitude but zero duration.

Initial velocity is $v_0 = I/M$ and motion stops when all the impulse has been dissipated by the retarding force, so that $T = I/P$. The acceleration of the mass is $a = -P/M$. Using Equation 5.5,

$$x_{max} = v_0 T + \tfrac{1}{2} a T^2$$

$$= \frac{PT}{M} T - \frac{PT^2}{2M}$$

$$= \frac{PT^2}{2M}$$

so that

$$T = \sqrt{\left(\frac{2 x_{max} M}{P} \right)} \tag{7.10}$$

(see Abrahamson and Lindberg, 1972).

Linearly increasing resistance
If the resisting force increases linearly with displacement the situation is exactly the same as for the linear undamped vibrating system discussed in Chapter 6. If all of the energy is dissipated during the first quarter-cycle, motion stops when the displacement reaches its first maximum value. Otherwise it may be necessary to use some ingenuity or trial and error.

Elastic/ideally-plastic resistance
If the resisting force increases linearly with displacement up to a yield value and then remains constant, the two previous analyses can be combined. It is convenient to define the yield by the yield deflection x_y.

If a constant force F is applied instantaneously to the mass, Equation 6.16 can be rewritten to give

$$t_y = \frac{1}{\omega} \cos^{-1} \left(1 - \frac{K x_y}{F} \right) \tag{7.11}$$

where t_y is the time at which yield occurs. By differentiation of Equation 6.16, the velocity of the mass at yield can be found to be

$$v_y = -\frac{F}{K} \omega \sin \omega t_y \tag{7.12}$$

After yield, Equations 7.3 and 7.4 apply, giving

$$v = v_y + \frac{F - P}{M} (t - t_y) \tag{7.13}$$

and

$$x = x_y + v_y (t - t_y) + \frac{F - P}{2M} (t - t_y)^2 \tag{7.14}$$

The maximum deflection x_{max} occurs when $v = 0$. If F is greater than P, deflection increases without limit. If F is less than P, x_{max} is finite.

Substituting $v = 0$ in Equation 7.13 gives

$$t_{max} = t_y \left(1 + \frac{M}{P - F} \right) \qquad (7.15)$$

and substituting this in Equation 7.14 gives x_{max} if motion stops once x_{max} is reached.

If the linearly increasing element acts as an undamped spring the situation is more complicated and the mass vibrates freely about a displaced mean position, as shown in Fig. 7.2. In this case x_{max} is reached each cycle and it can be shown that Equation 7.15 becomes

$$t_{max} = t_y + \frac{M}{P - F} \frac{F\omega}{K} \sin \omega t_y \qquad (7.16)$$

As soon as x_{max} is reached for the first time the retarding force P no longer acts but the applied step force F remains. This is equivalent to the instantaneous removal of a force $P - F$, and the differential equation of motion becomes

$$Ma + Kx = P - F \qquad (7.17)$$

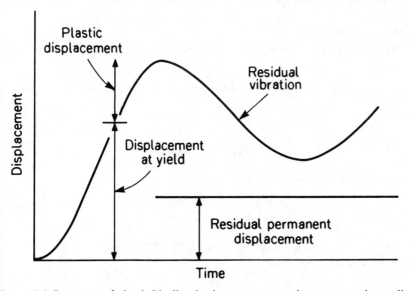

Figure 7.2 Response of elastic/ideally-plastic system to an instantaneously applied constant force.

The results of Section 6.2.2 apply, giving

$$x = x_{max} - \frac{P - F}{K}[1 + \cos \omega(t - t_{max})] \qquad (7.18)$$

This is a simple harmonic vibration of amplitude $\pm (P - F)/K$ about a mean position $x_{max} - (P - F)/K$.

The mean spring deflection caused by the constant load is F/K, and so the permanent deflection of the constant force element is

$$x_D = \left(x_{max} - \frac{P - F}{K} \right) - \frac{F}{K}$$

$$= x_{max} - \frac{P}{K} \qquad (7.19)$$

If the applied force increases linearly with time, reaching a steady value F_{max} at time t_d, its behaviour can again be treated in the same way as in Section 6.2.2(a). There are now five independent variables: pulse rise time t, natural period of the spring mass system T, applied force F, damper force P and yield displacement x_y. Even if results are restricted to maximum and yield amplitudes there are also three dependent variables, x_{max}, t_y and t_{max}, and so presentation of results can be difficult (Biggs, 1964).

Treatment of the response to simple pulses is substantially the same as that of step inputs and is based on Section 6.2.2. Because there are a number of variables this is usually confined to the numerical analysis of specific cases.

Viscous resistance

If the resisting force varies linearly with velocity, the differential equation of motion is

$$Ma + Cv = F \qquad (7.20)$$

giving

$$-\frac{C}{M}dt = \frac{dv}{v - (F/C)}$$

Integrating both sides gives

$$-\frac{C}{M}t = \log_e \left(v - \frac{F}{C} \right) + \log_e A \qquad (7.21)$$

where A is a constant which depends on the initial conditions.

If $v = 0$ when $t = 0$, then

$$0 = \log_e \left(-\frac{F}{C} \right) + \log_e A$$

$$= \log_e\left(\frac{-AF}{C}\right) \tag{7.22}$$

Taking antilogarithms gives

$$1 = -\frac{AF}{C}$$

or

$$A = -\frac{C}{F}$$

Then

$$-\frac{C}{M}t = \log_e\left(v - \frac{F}{C}\right)\left(-\frac{C}{F}\right)$$

$$= \log_e\left(1 - \frac{C}{F}v\right) \tag{7.23}$$

Taking antilogarithms gives

$$e^{-(C/M)t} = 1 - \frac{C}{F}v \tag{7.24}$$

or

$$v = \frac{F}{C}(1 - e^{-(C/M)t}) \tag{7.25}$$

Integrating gives

$$x = \frac{F}{C}\left(t + \frac{M}{C}e^{-(C/M)t}\right) + c \tag{7.26}$$

If $x = 0$ at $t = 0$, then the constant

$$c = -\frac{FM}{C^2}e^{-(C/M)t}$$

If F is a constant force applied instantaneously as a step when $t = 0$, then Equations 7.25 and 7.26 give the results shown in Fig. 7.3. v approaches a limiting value of

$$v_1 = \frac{F}{C} \tag{7.27}$$

If the mass is travelling at an initial velocity v_0 and only the resisting force acts on it, then the differential equation of motion is

$$Ma + v = 0 \tag{7.28}$$

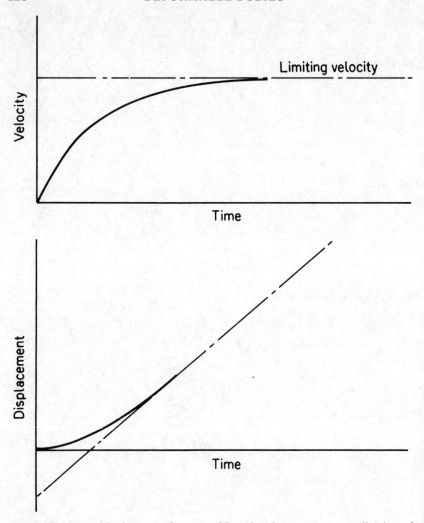

Figure 7.3 Mass with viscous resistance subjected to instantaneous application of a constant force.

giving

$$-\frac{C}{M}\mathrm{d}t = \frac{\mathrm{d}v}{v}$$

$$v = v_0 \mathrm{e}^{-(C/M)t} \tag{7.29}$$

and

$$x = \frac{Mv_0}{C}(1 - \mathrm{e}^{-(C/M)t}) \tag{7.30}$$

giving a limiting displacement of

$$x_{max} = \frac{Mv_0}{C}$$

For a rectangular force–time pulse of duration t_d, Equations 7.25 and 7.26 apply up to time t_d and then Equations 7.29 and 7.30 apply. As with a constant resisting force there is no characteristic response time, but the response to a pure impulse can be used to obtain a reference time. This gives

$$v_0 = \frac{I}{M}$$

and

$$v = \frac{I}{M}e^{-(C/M)t} \tag{7.31}$$

The velocity takes an infinite time to fall to zero, so an arbitrary reference velocity v_r needs to be specified. Often this is taken as $v_0/2$. From Equation 7.31,

$$v_r = \frac{I}{M}e^{-(C/M)t}$$

where T is the reference time, giving (Flügge, 1967)

$$T = -\frac{M}{C}\log_e\frac{Mv_r}{I} \tag{7.32}$$

(b) Damped vibrations

Free and forced vibrations
The simple model of a damped vibrating system is shown in Fig. 7.1(a). The spring and damper are in parallel and usually viscous damping is assumed. Other types of damping may be more realistic but are more difficult to deal with and the broad aspects of behaviour are substantially the same. Results are similar to those for undamped vibration but appreciably more complicated, and only simple comparisons between damped and undamped systems are given here. With viscous damping the differential equation of motion is for free vibration

$$Ma + Cv + Kx = 0 \tag{7.33}$$

where C is a damping constant with units of $N\,m^{-1}\,s^{-1}$.
  The natural frequency is

$$\omega_d = \sqrt{\left[\omega^2 - \left(\frac{C}{2M}\right)^2\right]} \tag{7.34}$$

where ω is the natural frequency of the system when the damping is absent. The natural frequency reduces with the amount of damping, becoming zero when

$$C = C_{\text{crit}} = 2M\omega_{\text{d}}$$
$$= 2\sqrt{(KM)} \qquad (7.35)$$

This value is known as *critical damping*, and the amount of damping present is usually stated as a proportion of the critical value. When damping is above the critical value, free vibration does not occur and the behaviour is viscous. Much of the damping found in metal structures is about $0.1\,C_{\text{crit}}$, and this has a negligible effect on the natural frequency. With high damping materials damping can be the critical value or greater. A damped free vibration eventually dies away, each displacement peak being lower than the one in the previous cycle. If an initial velocity v_0 occurs at displacement $x = 0$ and time $t = 0$, the envelope of the peaks decays exponentially with time, giving

$$\log_e\left(\frac{x_{1\max}}{x_{2\max}}\right) = \frac{C(t_2 - t_1)}{2M} \qquad (7.36)$$

where $x_{1\max}$ is the maximum displacement at t_1 and $x_{2\max}$ is the maximum displacement at t_2. Equation 6.3 becomes

$$x = \left[\frac{v_0 + (C/2M)x_0}{\omega_{\text{d}}}\sin\omega_{\text{d}}t + x_0\cos\omega_{\text{d}}t\right]e^{-(C/M)t} \qquad (7.37)$$

If the mass is subjected to a continuous sinusoidally varying force $F\sin\omega_1 t$, behaviour is analogous to undamped behaviour. There is a transient component which dies away, leaving a steady-state component with

$$x_{\text{abs}} = \frac{F/K}{\sqrt{\{[1 - (\omega_1^2/\omega^2)]^2 + (C^2\omega_1^2/M^2\omega^4)\}}} \qquad (7.38)$$

If the input is a continuous displacement $x_s\sin\omega_1 t$ at the base of the spring away from the mass, the damping force is usually assumed to be proportional to the relative velocity between the base and the mass. This gives a steady-state component with

$$x_{\text{r}} = \frac{x_s(\omega_1/\omega)^2}{\sqrt{\{[1 - (\omega_1^2/\omega^2)]^2 + (C^2\omega_1^2/M^2\omega^4)\}}} \qquad (7.39)$$

The spring distortion is related to the square of the exciting frequency in much the same way as a spring–mass system responds to an out-of-balance centrifugal force applied to the mass, where the exciting force increases with the square of the frequency.

Special analysis techniques

It is evident from the previous section that even very simple analysis can become complicated when a velocity-dependent term is present. For a single vibrating system the phase-plane method discussed in Section 6.2.1(c) can be used. Instead of a circle on rectangular axes the plot of x against v/ω becomes a logarithmic spiral on oblique axes. The amount of deviation from a circle is proportional to the damping.

The most widely used method is Laplace transforms, which convert differential and integral equations into algebraic equations, broadly as logarithms convert multiplication and division into addition and subtraction. As with logarithms, standard tables of transforms are available for many cases and the theory behind them need only be used when the standard tables are not suitable.

There are two main components of differential and integral equations which need to be transformed – the parameters, and the processes of differentiation and integration. A parameter X which is a function of time t can be written as $X(t)$ and its Laplace transform written as

$$\mathscr{L}[X(t)] = x(s)$$

The transformed parameter x is a function of s which is a new variable replacing t.

The process of differentiation has the transform

$$\mathscr{L}\left[\frac{\mathrm{d}X}{\mathrm{d}t}\right] = sx - X(0)$$

where $X(0)$ is the value of X at $t = 0$.

The differential equation of motion (Equation 6.1)

$$Ma + Kx = 0$$

can be rewritten

$$\frac{\mathrm{d}^2 X}{\mathrm{d}t^2} + \frac{KX}{M} = 0 \tag{7.40}$$

If the displacement is $X = A$ at $t = 0$ the differential equation transforms into the algebraic equation

$$(s^2 x - sA - 0) + \frac{Kx}{M} = 0 \tag{7.41}$$

where the term in brackets is the transform of the double differentiation and x is the transform of X. This gives

$$x = \frac{sA}{s^2 + (K/M)} \tag{7.42}$$

which transforms back into

$$X = A \cos \sqrt{\left(\frac{K}{M}\right)} t \tag{7.43}$$

$$= A \cos \omega t$$

where $\omega = \sqrt{(K/M)}$.

When an external force–time pulse is applied the transform for the pulse is used. For instance, the transform of an instantaneous step load of amplitude A applied at time $t = 0$ is A/s.

Laplace transforms are very useful in analysing the transient motion of damped vibrating systems, and published work contains many examples of their use. The nomenclature varies and the symbols used above, though typical, are not unique (Sensicle, 1968; Spiegel, 1965).

Response to simple pulses
The response of a damped vibrating system to any given simple pulse is broadly similar to that of an undamped vibrating system. Using impulse–momentum analysis, the convolution integral of Equation 6.11 becomes

$$x = \frac{1}{M\omega_d} \int_0^t F e^{-(C/2M)(t-\tau)} \sin \omega_d(t - \tau) d\tau \tag{7.44}$$

This can be combined with Equation 7.37 to give a comprehensive expression equivalent to Equation 6.13. Using Fourier analysis, the damped steady-state response is taken as the response function of the system.

Sometimes it is necessary to consider the response of the system in terms of the acceleration of the mass. If a sudden displacement is applied to the other end of the spring and the system is undamped,

$$a = \frac{1}{M}(F - Kx) \tag{7.45}$$

As long as the displacement x is finite, the acceleration a can be found.

If the system is viscously damped,

$$a = \frac{1}{M}(F - Cv - Kx) \tag{7.46}$$

If the displacement x is applied instantaneously, the initial velocity v_0 is infinite and the acceleration cannot be found. This precludes the use of displacement–time pulses with sudden discontinuities, and equivalent continuous pulse shapes are needed. These can be versed sine shapes, as discussed in Section 6.2.2(d), or rounded step functions of other types. For instance the equation

$$x = x_{max}[1 - e^{-\gamma\omega t}(1 + \gamma\omega t)] \tag{7.47}$$

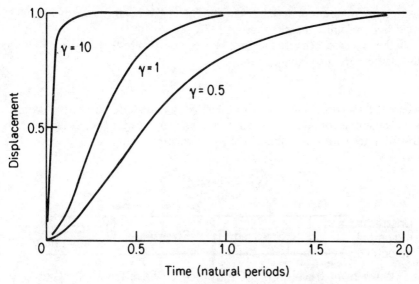

Figure 7.4 Typical rounded step displacements.

represents a step displacement starting at time $t = 0$. The natural frequency of the vibrating system is ω. The parameter γ is given by $T/2t_d$, where T is the natural period of the vibrating system and t_d is the time taken for the rounded step to reach a specified proportion of its maximum value. Figure 7.4 shows some typical rounded steps. In these the step reaches about 80% of its maximum value in time t_d. For other proportions the values of γ would vary but the general pattern of behaviour would remain the same.

Use of this type of more complicated step is greatly facilitated by Laplace transforms (Snowdon, 1968).

7.2.2 Stress waves

When the duration of the impact is comparable with, or less than, the time taken for a stress wave to traverse the body, stress waves become of interest. Stress waves are discussed in Chapters 1, 2 and 3 mainly in terms of infinitely long rods striking rigid anvils. If the impacting bodies are of finite size and are both deformable, and if the impact lasts for a finite time, further effects need to be taken into account. These are discussed below in terms of one-dimensional rods of finite length.

(a) Loading waves

Two cases of a one-dimensional stress wave moving along a rod are discussed: a moving bar squarely striking the end of a stationary rod of identical cross-section but different material; and a stationary bar being struck by a moving rigid body of finite mass (Johnson, 1972, pp. 15–16, 33–4).

In the first case, because the force between the two rods is common to both, the stresses are also equal, giving

$$\sigma_1 = \sigma_2$$

where the suffixes 1 and 2 refer to the moving and stationary rods. At time t after the initial contact the situation is as shown in Fig. 7.5(a), using Eulerian co-ordinates with the origin at the point where contact was first made. The

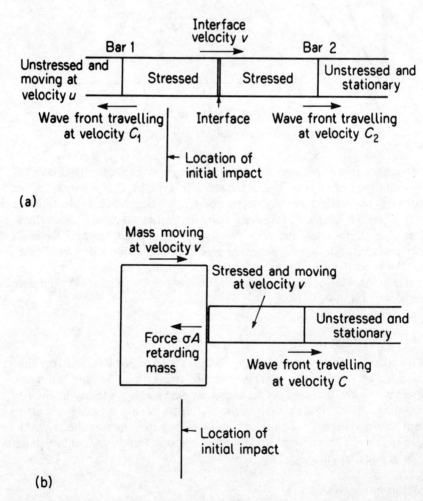

(a)

(b)

Figure 7.5 (a) Impact of two rods of identical cross-section but different materials. (b) Impact of stationary rod and moving rigid mass.

interface between the two bars is moving in the same direction as the moving bar at velocity v, giving

$$\sigma_1 = \rho_1 C_1 (u - v)$$

$$\sigma_2 = \rho_2 C_2 v$$

where u is the original velocity of the moving bar. It is useful to introduce the concept of *mechanical impedance*, which is

$$\rho C = \frac{\sigma}{v} = I \qquad (7.48)$$

Then

$$I_1(u - v) = I_2 v$$

giving

$$v = \frac{I_1 u}{I_1 + I_2}$$

It follows that

$$\sigma_1 = I_1 (u - v)$$

$$= I_1 u \left(1 - \frac{I_1}{I_1 + I_2} \right)$$

$$= \frac{I_1 I_2 u}{I_1 + I_2}$$

$$= I_2 v$$

$$= \sigma_2$$

In the second case the impulse applied to the moving mass by the rod equals the change in momentum of the mass. Thus if its velocity at any time t is v, then

$$M \frac{dv}{dt} = -\sigma A$$

where A is the cross-sectional area of the bar. The situation at time t is shown in Fig. 7.5(b). It follows that

$$M \frac{dv}{dt} = -IAv$$

or

$$\frac{M}{A} \frac{dv}{v} = -I\, dt$$

Integrating gives

$$\frac{M}{A} \int_u^v \frac{dv}{v} = -I \int_0^t dt$$

where u is the initial velocity of the mass. The solution to this equation is

$$v = u\,e^{-(AI/M)t} \tag{7.49}$$

Alternatively,

$$\sigma = Iu\,e^{-(AI/M)t}$$

This stress decreases exponentially with time.

(b) Unloading waves

If a stress pulse is of finite duration, it begins with a loading wave which imposes the stress and finishes with an unloading wave which removes the stress. Unloading waves are generally assumed to travel at the linearly elastic wave velocity; hence, if the stress pulse is linearly elastic, it travels with its shape unchanged because all parts of it travel at the same velocity.

If the stress pulse is non-linear, with E_{tan} decreasing with increase in stress, the loading wave disperses as it travels, with the larger stresses travelling more slowly than the smaller ones. In this case the unloading wave catches up with the larger stresses and cancels them. It never catches the initial, linearly elastic, part of the stress pulse because they both travel at the same velocity. This behaviour is shown in Fig. 7.6(a).

If the stress pulse is non-linear, with E_{tan} increasing with increase in stress, the pulse sharpens up to a shock wave as discussed in Section 2.4.1(c). In this case the unloading wave never catches up with the shock wave, which decays, however, because of energy losses at the wave front. Any residual pulse left is likely to travel at the linearly elastic velocity. This behaviour is shown in Fig. 7.6(b) (Johnson, 1972, pp. 218–19).

(c) Reflected waves

If a simple wave of direct elastic stress travels along a rod of finite length, it eventually reaches the end remote from the impact end. There it is reflected back along the rod, still at the elastic wave velocity. There are two limiting conditions. In the first the remote end of the rod is completely unrestrained, and in the second it is rigidly fixed.

If the end is unconstrained, no stress can remain there, and so the reflected wave must unload the rod. This is equivalent to superimposing a stress wave on top of the loading wave which is equal in amplitude to it but opposite in sign.

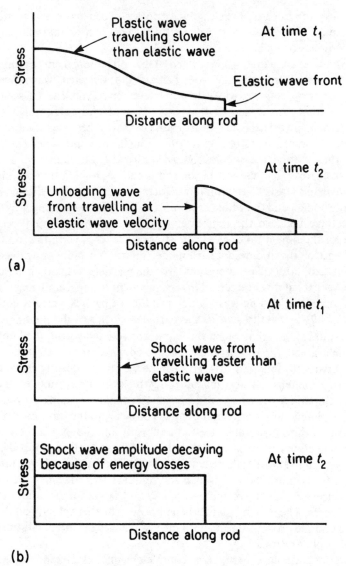

Figure 7.6 Unloading of stress waves in a rod. (a) Elastic unloading wave. (b) Shock wave decaying.

With a compressive wave, the particle velocity is in the same direction as the wave propagation because the impacted end moves in the same direction as the wave front. With a tensile wave it is in the opposite direction because the impacted end moves away from the wave front.

When a compressive wave is reflected as a tensile wave the wave front changes direction but the particle velocity does not, because it is now moving in the opposite direction to the wave front. This means that the particle velocity doubles on reflection.

The overall result is that, when the end of the rod remote from the impact is completely unconstrained, the rod behind the reflected wave front is unstressed but moving at twice the impact velocity. This can be seen in a straight rod impacting a rigid anvil at a velocity v which is low enough for the stress to remain linearly elastic. The front end of the rod becomes stationary on impact and a compressive stress wave runs along the rod until it reaches the far end. At this instant the whole rod is stationary and under a uniform compressive stress. The far end of the rod is then unloaded by the reflected tensile wave and starts to rebound at velocity $-v$. When the reflected wave reaches the impact end the whole rod is unstressed and rebounding at velocity $-v$. The force between the impact end of the rod and the anvil remains constant for the time it takes an elastic stress wave to traverse the rod twice.

If the end of the rod remote from the impact is rigidly restrained, no movement can take place there, and so the particle velocity behind the reflected wave front must be zero. This is equivalent to superimposing a stress wave on top of the loading wave which produces a particle velocity equal in amplitude but opposite in sign to the particle velocity of the loading wave.

The overall result is that, when the end of the rod away from the impact is rigidly constrained, the rod behind the reflected wave front is stationary but stressed at twice the impact stress. This can be seen in a rod being compressed between rigid platens in a test machine, in which one platen moves and the other is stationary. A stress wave of amplitude σ propagates along the rod from the moving platen until it reaches the end in contact with the stationary platen. At this instant the entire rod is stressed uniformly and moving at the velocity of the moving platen. Movement of the far end of the rod is then stopped by the stationary platen and a reflected wave of a further amplitude σ starts to propagate back along the rod. When the reflected wave reaches the end in contact with the moving platen the whole rod is stationary and stressed uniformly at 2σ. The moving platen then imposes a further velocity on the end of the rod and the process repeats until the rod reaches the intended stress and the moving platen stops.

An intermediate case between a completely unconstrained and a fully constrained end is when the end of the rod is attached to a rigid mass of magnitude M. In this case the analysis given in Section 7.2.2(a) is used with the relationship

$$v = u\,\mathrm{e}^{-(AI/M)t}$$

where u is the particle velocity of the loading wave and v is the velocity of the mass at time t after the front of the loading wave reaches the end of the rod.

The stress in the reflected wave decays exponentially with time, but the particle velocity increases as the velocity of the mass decreases until particle velocity and mass velocity are equal.

If the loading wave is also caused by a finite rigid mass, the wave reflects between both ends of the rod until an equilibrium state is reached. For some lengths of rod both masses and the rod will then be moving at a common velocity with the rod unstressed. However, for most lengths of rod there will be a residual stress wave oscillating up and down the rod, because the time when a mass is stationary rarely matches the time when the stress wave reaching it produces a net force of zero. This is analogous to the behaviour of a single-mass vibrating system subjected to rectangular pulses of different durations.

Some materials are appreciably weaker in tension than in compression, and in addition it is possible to transmit a compressive load across a crack but not a tensile one. This leads to a phenomenon known as *spalling*, which occurs when a compressive loading wave is reflected at an unconstrained surface as a tensile unloading wave. The material can then fracture in tension and the fractured part flies off. The exact location of the fracture can be hard to predict because it is likely to be located at a weak point in the material.

7.3 MORE COMPLEX ANALYSIS

There are two common extensions of the analysis of a simple deformable body. These occur when the load deflection characteristics of the body are appreciably non-linear, and when behaviour is analysed in two or three dimensions. Because of the number of variables involved, published analysis uses either sophisticated mathematics leading to complex general relationships or numerical methods leading to results for specific cases.

Extension to two or three dimensions can produce distinct changes in stress wave patterns. For instance, a compressive stress wave reflecting from a free surface at an angle produces not only a reflected tensile wave but also a shear wave and a wave along the surface.

8

Energy absorbers

8.1 INTRODUCTION

8.1.1 General

(a) Energy absorption

Energy is absorbed on impact in a limited number of ways. Deformation of solids is the most common; usually energy is absorbed in plastic flow, although appreciable amounts of energy can also be absorbed in controlled brittle fracture. Fluid flow can also be used, usually with the fluid being forced through an orifice. Friction is always present to some degree. However, coefficients of friction change unpredictably and thus, unless some subsidiary correcting system is used, absorbers relying solely on friction are only poorly repeatable. In most cases the behaviour of impact absorbers can be predicted approximately, but accurate performance figures need to be found by trial and error.

Absorbers can be *reusable*, as in a hydraulic damper; *rechargeable*, with the energy absorbing component being replaced in a permanent container; or *expendable*, as in the collapse of a vehicle structure during a crash.

The energy absorber used depends on the nature of the impact load. This can be distributed over the whole of the impacted body, as in blast loading, or it can be localized, with a small or pointed body hitting a large one. The large body may deform overall in broadly the same way as if the load were distributed, or the small body may penetrate it locally. Usually the overall response is appreciably slower than the localized response and so they can be analysed separately, but sometimes the two types of deformation respond at similar rates and need to be analysed simultaneously.

(b) Impact speeds

Impact speeds of general interest for energy dissipation fall very approximately into three main ranges, as can be seen in Fig. 8.1. At speeds from, say, 1 to $15 \, \text{m s}^{-1}$, impact behaviour can be related to static or vibration behaviour. Large objects with weights measured in tonnes can probably be stopped without departing significantly from normal methods of construction. This

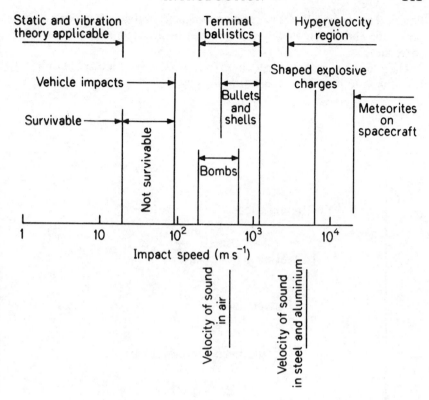

Figure 8.1 Impact speeds for energy absorbers.

range includes survivable vehicle accidents in which the aim is to stop a crashing vehicle without damaging or releasing its contents.

At speeds from, say, 150 to $1500\,\mathrm{m\,s^{-1}}$, dynamic effects make impact behaviour appreciably different from static behaviour. However, material properties can probably be related to their static properties, although strain rate effects are important. This is approximately the range covered by the study of conventional terminal ballistics with bombs, bullets and shells. These are regarded as dense rigid objects, weighing from a few grams to a few hundred kilograms, which need to be stopped without excessive damage to the target. There are also problems, such as an aircraft crashing at full speed into a nuclear power plant, where the aircraft is regarded as a deformable object which needs to be destroyed. Specialized forms of construction are needed, involving armour or special reinforcement. These are usually much heavier than normal forms of construction but can be related to them.

At speeds from, say, 3000 to $30000\,\mathrm{m\,s^{-1}}$, behaviour is completely unlike static behaviour in all respects. Materials are vaporized and solids flow as liquids. The range includes jets of liquid metal from shaped explosive charges penetrating heavy armour, and micrometeorites impacting lightly constructed spacecraft.

The intermediate speed ranges of 15 to 150 m s^{-1} and 1500 to 3000 m s^{-1} do not appear to have been studied extensively.

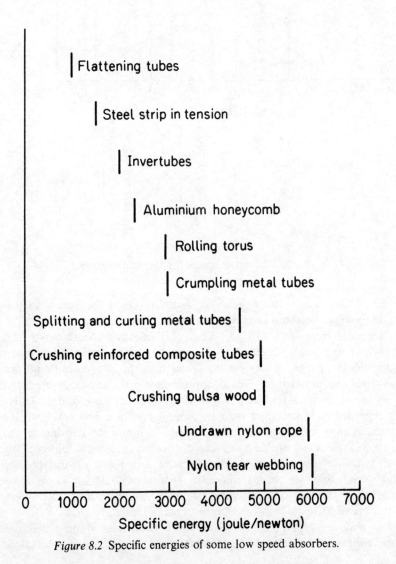

Figure 8.2 Specific energies of some low speed absorbers.

8.1.2 Basic relationships

(a) Efficiency

The efficiency of an energy absorber can be specified in a number of ways. Two widely used ones are *specific energy*, which is the energy absorbed per unit weight, and *volumetric efficiency*, which is the percentage of the total volume of the absorber which is employed usefully. The latter is often simplified to *stroke efficiency*, which is the length of stroke divided by the total length of the absorber. It is also possible to measure efficiency in terms of an ideal energy absorber. The characteristics of this vary with the dynamics of the system, but as a general comparator an *ideal absorber* is usually taken to be one which instantaneously reaches the maximum allowable retarding force and retains this force throughout the stroke, giving a rectangular force–time pulse.

Often it is possible to incorporate energy absorbing properties into a component which has another function and, in this case, efficiency becomes less important. A common example is a structural member which is designed to collapse in a controlled manner, absorbing a significant amount of energy. This can be less than optimum as a structural member and relatively inefficient as an energy absorber, but it may be the best overall compromise.

Some typical efficiencies are shown in Fig. 8.2.

(b) Reliability and versatility

An important requirement of an energy absorber is that it should be reliable. Sometimes an energy absorber (or a set of replaceable absorbers) is in constant use and so its performance can be monitored, but often an absorber is there only for use in an emergency. It may remain for a long time neglected and forgotten but need to function properly when the need arises.

Versatility is also important because more than one type of impact may be possible and these may occur at a variety of angles. It is difficult to give an objective measurement of versatility and, though reliability can be specified objectively, it is frequently given in broad subjective terms. Reliability and versatility are often in conflict with efficiency.

8.2 ISOLATING MOUNTINGS

8.2.1 Steady-state vibration

(a) Requirements of an isolating mounting

The purpose of a vibration isolating mounting is to reduce the transmission of vibrations. This can be demonstrated by considering the response to a constant amplitude sinusoidal impact of the simple undamped and damped vibrating systems discussed in Sections 6.2.2(a) and 7.2.1(b). Well below the natural frequency, the output is the same as the input. Close to the natural frequency, the output is greater than the input by an amount which depends on how much damping is present. Above the natural frequency, the output falls progressively below the input but falls more quickly with the undamped system than with the damped one.

To reduce the response throughout the frequency range of interest, it is desirable to have a low natural frequency, so that most of the frequencies of interest are well above it, and to have sufficient damping to keep the amplification at the natural frequency within acceptable limits.

The performance of a vibration isolating mounting can be measured by its *transmissibility*. This is usually specified in two different ways, which can be shown to be identical for linear behaviour. In the first the end of the spring away from the mass is subjected to a sinusoidal displacement of constant peak amplitude, and the transmissibility is taken as the maximum displacement of the mass divided by the maximum displacement of the input. In the second the end of the spring away from the mass is rigidly fixed and a sinusoidal force of constant peak amplitude is applied to the mass. The transmissibility is the maximum force at the fixed end divided by the maximum force applied to the mass (Snowdon, 1968, pp. 21–3).

(b) Three-element mounting

Performance can be improved by adopting a three-element mounting as shown in Fig. 8.3. In this there are two springs in parallel plus a viscous damper in series with one of the springs. At low frequencies the resistance of the damper is small and the spring in series with it is not loaded. At high frequencies the resistance of the damper is large and both springs are deflected by the same amount. In both cases the energy lost in the damper is small. At intermediate frequencies the overall spring stiffness changes with frequency. The damping is significant and changes with frequency, reaching a maximum value at a transition frequency and then falling again. Maximum energy loss through damping occurs when the transition frequency is equal to the natural frequency of the overall vibrating system. Behaviour for a three-element

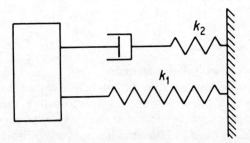

Figure 8.3 Three-element absorber.

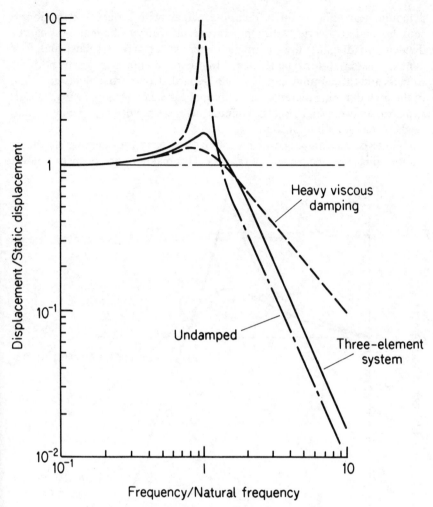

Figure 8.4 Transmissibility of typical three-element system.

mounting is shown in Fig. 8.4, and is compared with an undamped system and one with a single spring in parallel with a viscous damper. With the three-element mounting it is possible to reduce transmissibility at resonance and also to have a rapid reduction in transmissibility with frequency above resonance (Snowdon, 1968, pp. 33–8).

(c) Rubber mountings
Anti-vibration mountings are often made from rubber-like materials. These provide both spring and damping and are usually loaded in shear. Both the

dynamic shear stiffness and the damping increase with frequency. Behaviour is complex and also varies with temperature, but rubber-like materials can be divided broadly into low damping materials (or type I solids) and high damping materials (or type II solids). In the low damping materials both the stiffness and the damping can usually be regarded as approximately constant. In the high damping materials the damping can again usually be regarded as approximately constant, but the stiffness increases rapidly with frequency and needs to be taken into account.

Unlike viscous damping, which varies with velocity, the damping of rubbers is approximately independent of velocity. This makes it inconvenient to use a

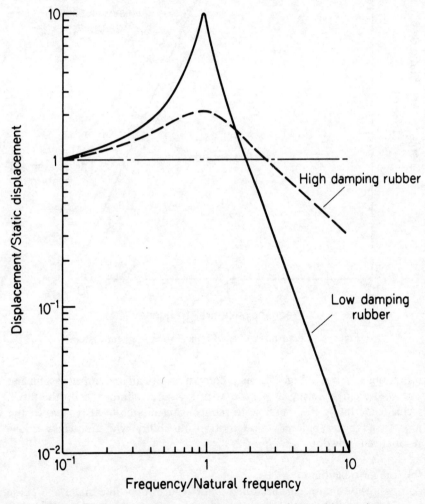

Figure 8.5 Transmissibility of a simple system with a rubber spring.

damping ratio which specifies damping as a proportion of critical damping. Instead a damping factor is used which is defined in terms of the ratio of the imaginary to the real part of the dynamic shear modulus when this is expressed as a complex number. Typical damping factors are about 0.1 for low damping and 1.0 for high damping.

The transmissibility of simple single-mass vibrating systems is shown in Fig. 8.5. In these both spring and damper are provided by rubber in shear (Snowdon, 1968, Figs 2.2 and 2.3). With the low damping rubber, transmissibility decreases rapidly above the natural frequency. This is an effect of the velocity-independent damping. With the high damping rubber, two effects occur: the velocity-independent damping reduces the transmissibility above the natural frequency, but the increasing spring stiffness increases it. These tend to cancel out, and so the system with the high damping rubber behaves in much the same way as a system with viscous damping. Transmissibility decreases above the natural frequency but more slowly than with the low damping rubber.

Performance can be modified by using low and high damping rubber in parallel or in series. Analysis becomes rather complicated and the design of such mountings is largely empirical (Freakley and Payne, 1978, Chapter 7).

(d) Stress waves

If the frequency of vibration is high enough the time taken by a single cycle is comparable with the time taken by a stress wave in traversing the mounting. The effects of stress waves then become important because they increase the transmissibility. Standard stress wave analysis is used but its combination with vibration analysis leads to complex calculations. In addition the propagation of stress waves in a rubber mounting can be difficult to analyse, so problems are usually treated empirically (Freakley and Payne, 1978, pp. 83–93).

8.2.2 Impact loading

(a) Three-element mounting

The response of a three-element mounting with viscous damping is substantially similar to that of the simple damped system discussed in Chapter 7. It reduces the displacement in the same way as a heavily damped simple system but without the corresponding increase in acceleration found in the simple system (Snowdon, 1968, pp. 397–400).

(b) Rubber mats and mountings

It is not always easy to distinguish between a rubber anti-vibration mounting and a shock absorbing one. Sometimes the same unit serves both functions but sometimes the requirements are different, although it is still necessary to

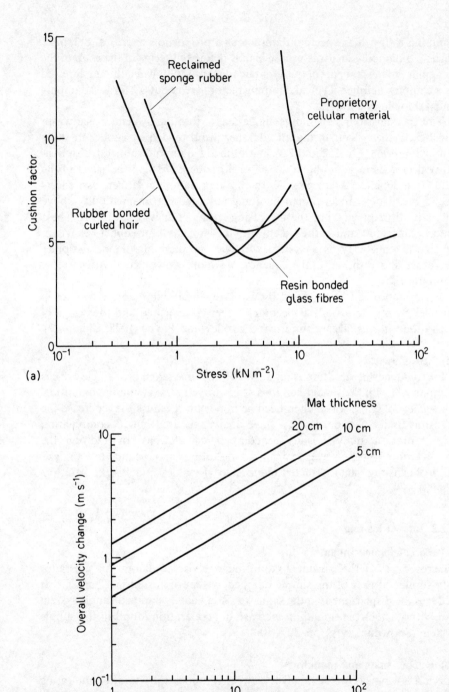

Figure 8.6 (a) Typical cushion factors for energy absorbing mats. (b) Some typical properties of packaging mats.

ensure that a unit used to solve one problem does not make the other problem worse.

The analysis of rubber-like materials used to cushion impact loads appears to be largely empirical. The material is often in the form of a mat which is loaded in compression. The performance of such a mat may be specified by the dimensionless *cushion factor*

$$C = \frac{\sigma}{V_1}$$

Where σ is the stress and V_1 is the strain energy per unit volume up to that stress. Cushion factors for some typical materials are shown in Fig. 8.6(a). The shapes of the cushion factor curves are very similar and they can be reduced approximately to a single curve by using the dimensionless values

$$\bar{\sigma} = \frac{\sigma}{\sigma_0} \qquad \bar{\varepsilon} = \frac{\varepsilon}{\varepsilon_0}$$

where σ_0 and ε_0 are the stress and strain at which the cushion factor is a minimum.

The thickness of cushioning material needed is given by

$$t \simeq \frac{Chg}{a_{max}}$$

where h is the height through which the mass is dropped and a_{max} is the maximum allowable acceleration.

Performance can be improved by shaping the mats in various ways or by using the material in the form of a foam. It is also possible to use mountings similar to rubber anti-vibration mountings with the rubber loaded in shear. The design of these appears to be largely empirical.

The impact velocities involved are low, being up to about $25 \, \text{m s}^{-1}$. Fig. 8.6(b) shows typical values of maximum accelerations transmitted by mats of different thicknesses for different velocity changes (Freakley and Payne, 1978, Chapter 9). The mats have been selected to give the most effective cushioning. The acceleration is given in multiples of g, the acceleration due to gravity at the earth's surface.

8.3 DEFORMING SOLIDS

8.3.1 Without penetration

(a) Tension

The simplest form of energy absorber uses the permanent deformation of a strip of material loaded in tension. This stretches until fracture occurs and, with ductile materials, appreciable amounts of energy can be absorbed. The

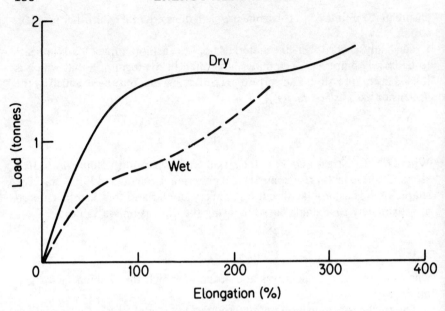

Figure 8.7 Typical properties of 7.5 cm circumference undrawn nylon rope.

absorber is insensitive to the direction of loading because it can be arranged so that it aligns itself with the direction of the load. Its main drawbacks are that failure is catastrophic and that there may be problems with localized failure of the material because of imperfections or because of stress wave effects.

With metals the material can be a rod, strip or tube. With textiles it can be a webbing strap or a rope. A useful textile material is undrawn nylon rope which can stretch up to 250% before failure, although there is significant elastic rebound. Typical stress–strain curves are shown in Fig. 8.7. No two ropes are identical, so the energy absorbed by two nominally equivalent ropes can vary by about 10%. Successful use of these ropes requires practice. Another useful textile material is tear webbing, which consists of two parallel strips of webbing partially woven together during manufacture. As one strip is torn away from the other energy is absorbed. Textile absorbers are useful for stopping moving bodies gently at moderately low velocities (Brabin, 1968).

(b) Compression

Axial
Axial compression absorbers are usually thin walled tubes, generally of circular cross-section. Good correlation exists between static and impact behaviour for aluminium tubes, although strain rate effects can increase the

impact loads for steel ones. The useful speed range appears to be up to, say, $30\,\mathrm{m\,s^{-1}}$. A major limitation is that the loading needs to be within a few degrees of axial.

In the simplest form a ductile tube is compressed between two parallel platens and allowed to crumple into its natural deformed pattern, which consists of a number of small local buckles. After an initial peak load, which starts the process, the buckles fold flat in succession, giving a load which fluctuates about a mean value as shown in Fig. 8.8(a). The initial peak load can be eliminated by prefailing the tube, and the static mean crumpling load for a tube of thickness t and diameter D is given approximately by

$$P = 6\sigma_y t\sqrt{(Dt)}$$

If the tube is more than a few diameters long it fails by bending overall rather than by crumpling, and absorbs much less energy. Variation in behaviour with length and wall thickness is shown in Fig. 8.8(b) (Andrews et al., 1983). Tubes of rectangular or square cross-section behave in a similar way but are less efficient and predictable. They are useful when the main role of a component is as a beam, with a secondary role as an energy absorber, and their performance can sometimes be improved by filling them with plastic foam which stabilizes the walls so that the local buckles form more effectively. The static mean crumpling load is given approximately by (Meng et al., 1983)

$$P = 16t^2 \sigma_y$$

Tubes of brittle fibre-reinforced composite material crush progressively to powder from one end, if the proportions are correct and one end of the tube is tapered to help start the crushing process. The load deflection curve is much flatter than for metal tubes, as shown in Fig. 8.8(c), and the specific energy can be much higher because of the low weight of the material (Hull, 1983).

By using a suitable die or by preforming the tube, it is possible to turn inside out a thin walled tube of circular section and very ductile material. The four main ways of doing this are shown in Fig. 8.9(a). Such tubes are known as *invertubes* and have smooth load–deflection curves as shown in Fig. 8.9(b). The inverting load for a tube of thickness t and diameter D is given approximately by (al Hassani et al., 1972)

$$P = 7.5\sigma_y t\sqrt{(Dt)}$$

The inverting process is quite sensitive to variations in the tube material and dimensions. Fig. 8.9(c) shows how the inverting load lies between the mean and the maximum crumpling load, so that if crumpling starts it is likely to take over from inverting (Searle and Brabin, 1970). Invertubes can be tapered slightly to give a gently rising load–deflection curve, tapered at both ends so that both ends invert simultaneously, placed one inside another to operate in

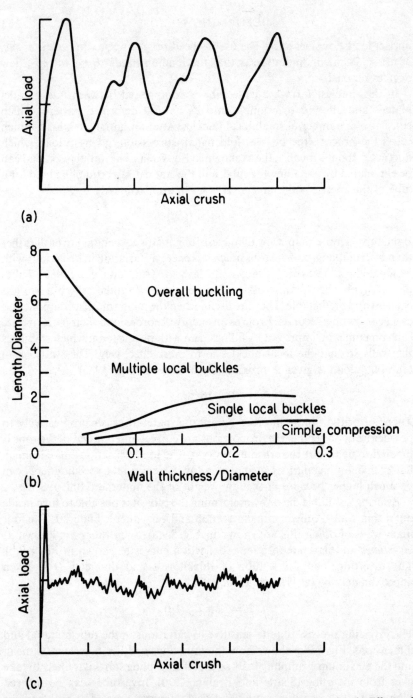

Figure 8.8 Axial loading of cylindrical tubes. (a) Thin walled metal tube. (b) Effects of dimensions of metal tubes. (c) Fibre-reinforced composite tube.

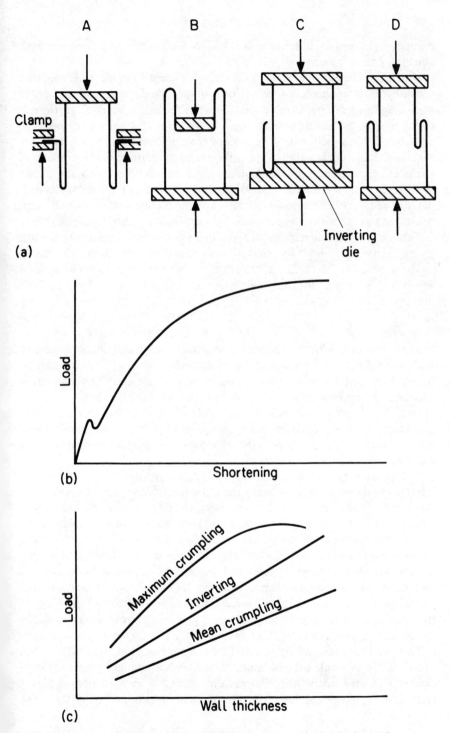

Figure 8.9 (a) Four types of invertube. (b) Typical invertube load–shortening curve.
(c) Inverting load in relation to crumpling loads.

parallel; and arranged in series with a double-sided inverting die between each adjacent pair of tubes (Sadeghi, 1984).

If an invertube with an inverting die is not sufficiently ductile it will split into a number of strips, each of which curls round at the die independently of the others. Splitting can be introduced deliberately by cutting short slits in the end of the tube or by having cutting edges on the die. This gives a curling tube absorber. It has a better stroke efficiency than an invertube, which can only shorten to half its original length; a curling tube can shorten to the diameter of a curl. Other differences are that square or rectangular tubes can be curled but not inverted, and less ductile material can be used. The curling load varies with the tube shape, the diameter of the curls and the friction, but needs to be kept below the load at which the tube fails by local buckling (Stronge *et al.*, 1983).

Tubes of very brittle material will fragment as they are forced over the die, giving behaviour roughly like that of reinforced composite tubes. The end of the tube needs to be chamfered to start the process. The fragmenting load depends on several factors but again needs to be kept below the load at which the tube starts to buckle.

Transverse

Energy absorbers which are loaded transversely are usually either sheets or slabs of material or thin walled tubes loaded transversely. Sheets can be rubber-like material or foamed plastic, in which case their performance is specified by a cushion factor as discussed in Section 8.2.2(b). They can also be of honeycomb, which consists of thin sheets of metal or paper, shaped and glued together to give a pattern of hexagonal cells like bees' honeycomb. There can be variations in the shapes of the cells, such as an eggbox pattern in which the cells are of square cross-section.

A honeycomb material can be regarded as a large number of thin walled tubes parallel to each other. Usually the tubes run through the thickness of the sheet so that, when it is loaded transversely, the tubes are loaded axially and crumple up in a pattern of small local buckles. Because of their number the load fluctuations cancel out, giving a flat load–deflection curve. Moreover, because they support each other, they do not bend overall, and so the honeycomb sheet can be very thick and is relatively insensitive to the angle of loading. At a loading angle of approximately 45° behaviour changes to that of a collection of thin walled tubes loaded transversely. It absorbs less energy in the latter case. Honeycomb manufacturers give energy absorbing properties for various grades; typically these range from 2.5 to $25 \, \mathrm{J \, cm^{-3}}$.

Thin walled tubes can be crushed transversely so that the cross-section becomes flat. A single tube of length L and with a circular cross-section of diameter D, crushed between flat parallel platens with reduction in platen separation δ, gives the load–deflection curve shown in Fig. 8.10(a). The

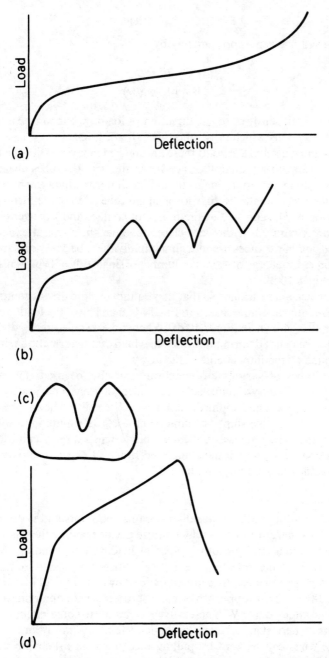

Figure 8.10 Tubes compressed transversely. (a) Single tube crushed transversely. (b) Three layers of crossed tubes crushed transversely. (c) Cross-section of crushed tube where sideways expansion was prevented. (d) Tube prevented from expanding sideways.

crushing load is given approximately by

$$P = \frac{2\sigma_y t^2 L}{D\sqrt{[1 - (\delta/D)^2]}}$$

Behaviour is not sensitive to the direction of loading, but can be altered by shaping the platens which crush the tubes (Reid, 1983).

Tubes can be used in arrays. If these are stacked in layers, with the tubes in alternate layers at right angles to each other and with the tubes in each layer about one diameter apart, deformation is uniform along each tube and behaviour is very similar to that for a single tube. If they are further apart, deformation is concentrated at the points of contact and the overall energy absorption is reduced. If they are closer together their lateral expansion is restrained, and if the tubes are touching initially the tube array can absorb up to three times the energy of a similar number of single tubes. Typical results are given in Fig. 8.10(b).

Tubes which are restrained so that they cannot expand do not compress flat but deform into the shape shown in Fig. 8.10(c) and have the load–deflection characteristics shown in Fig. 8.10(d). When tube arrays are loaded by an impact they may not deform uniformly, in which case deformation tends to be concentrated at the impact end of the array.

Rubber tubes of various cross-sections can also be used. The simplest example is an old car tyre being used as a fender between a boat and a jetty, but much more complex and sophisticated devices are in use. They are mainly for low speed impacts like ships berthing or motor cars running into obstacles when parking. These devices appear to be designed largely empirically by specialist companies which have acquired practical experience over several years (Freakley and Payne, Chapter 12).

(c) Bending

In general, bending occurs naturally in beams and struts and is discussed in Chapter 9. In energy absorbers which utilize plastic bending there is usually a mechanism which converts other types of loading into bending. A simple device is the W frame, shown in Fig. 8.11(a). When compressed from the ends this has the load–deflection characteristic shown in Fig. 8.11(b) (Johnson, 1972, pp. 184–5). There appear to be major problems with obtaining adequate joints at the angles of the W. These can be overcome to some extent by having an S frame, where sharp changes in angle at joints do not occur.

Ductile wires can be bent by pulling them through a four-pin device as shown in Fig. 8.12(a), or a similar device can be pushed along a wire which is fixed at each end. The wire is successively bent and straightened and the device produces a constant load. Friction can be reduced by having hardened rollers on the pins. The bending device is likely to be fairly heavy, but it can be

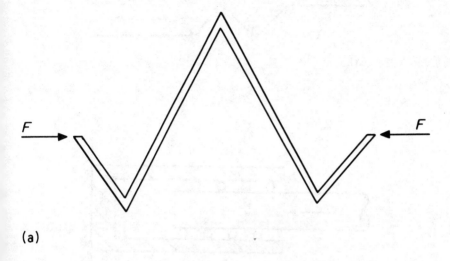

(a)

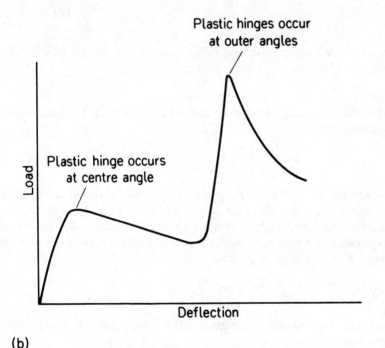

Plastic hinges occur
at outer angles

Plastic hinge occurs
at centre angle

Load

Deflection

(b)

Figure 8.11 W frame changing compression into bending. (a) W frame. (b) Load–deflection curve.

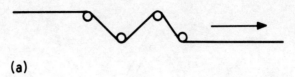

(a)

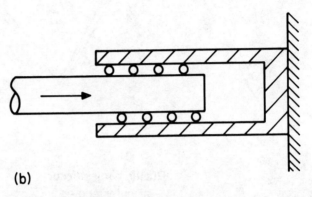

(b)

Figure 8.12 Wire bending devices. (a) Pulling wire through fixed rollers. (b) Rolling torus.

incorporated in a component which has another function. The load is given approximately by

$$P = \frac{4\sigma_y}{\sqrt{3}}\left\{ t - 2R\left[\sqrt{\left(1 + \frac{t}{R}\right)} - 1 \right] \right\}$$

for each roller, where R is the roller diameter (Johnson and Mamalis, 1978).

Another device is the rolling torus shown in Fig. 8.12(b). A spiral of wire of circular cross-section fits tightly between an inner mandrel and an outer tube. As the mandrel moves axially within the tube, the wire is continuously rotated and bent. The load is given approximately by

$$P = \frac{4}{3}\pi r^2 \sigma_y$$

for each ring of wire of radius r (Johnson, 1973). The radius of the ring itself is of no importance.

(d) Other
Energy absorbers using ductile materials can function in a variety of other ways and a large number of devices have been suggested or used. Often these

are based on conventional metal working processes such as cutting and pressing, but with the tools deliberately designed to absorb as much energy as possible. Because energy absorbers have to be designed semi-empirically it is usually more convenient to use existing types of device for which the empirical parameters are known.

8.3.2 With penetration

(a) Penetration and perforation
Often, when two bodies collide, one penetrates the other. This can occur over the entire range of impact speeds from static loading to hypervelocity impact. The penetrating body may be much smaller than the penetrated body or about the same size as it, and can range from relatively undeformable to relatively soft. The processes involved are usually very complicated and are not well understood, so empirical relationships and simple models are widely used. In general the study of penetration is aimed at limiting the damage caused to the penetrated body. Often this is done by placing some sort of protective layer in front of it.

The subject has been of interest in military applications for a long time and, more recently, it has become of interest in non-military applications as well. Penetration can occur, for instance, when vehicles crash, when high speed rotating machinery disintegrates and when pressure vessels explode.

Perforation occurs when one body makes a hole completely through the other, or through some element of it such as an outer skin. The transition from penetration to perforation is important and is specified, for instance, by the *ballistic limit*, which is the impact velocity at which perforation just occurs, or the *just contained energy*, which is the kinetic energy at which perforation just occurs. There are different definitions of the conditions at which perforation just occurs. For a metal plate these range from the nose of the impactor just breaking through the back face, to the impactor as a whole passing through the plate. Whichever definition is used there is usually considerable scatter in empirically observed ballistic limits.

(b) Targets and projectiles
Military terms are used frequently, even in non-military applications, with the penetrated body being called the target and the penetrating body the projectile or the bullet. Behaviour depends on the characteristics of both target and projectile as well as on the impact speed. Targets can be divided roughly into ductile and frangible and the difference between the two is appreciable. Ductile targets are usually of metal and frangible ones are usually of soil or concrete. Fibre-reinforced composites do not appear to be widely used as energy absorbers. The target can be an item of direct interest or it can be a protective layer such as an armour plate or compacted soil.

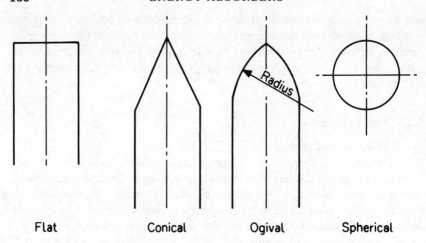

Figure 8.13 Typical nose shapes of non-deforming projectiles.

Projectiles can be divided roughly into deforming and non-deforming. Usually in designing protective layers it is assumed that the projectile is non-deforming. This greatly simplifies the analysis, and gives conservative results because when a projectile deforms there is usually a marked reduction in penetration. Non-deforming projectiles are generally assumed to be either spherical or cylindrical with a nose of one of the shapes shown in Fig. 8.13. As well as its shape the mass of the projectile is important, and this is often presented in the form of a density. Because many of the relationships used are empirical, the weight can be used instead of the mass. For instance, calibre density is given as $= W/D^3$, where W is the weight of the projectile and D is its diameter (see later). On this definition a long projectile is more dense than a short one of the same diameter and material.

If the projectile deformation needs to be taken into account, analysis becomes appreciably more complex because of the greater number of variables. Empirical or semi-empirical relationships can still be established but overall patterns of behaviour do not seem to have emerged. In theoretical analysis the changing interface between the bodies needs to be allowed for and numerical calculations can become very lengthy.

(c) Metal plates
Analysis of the penetration of metal plates is usually restricted to a non-deformable projectile striking the plate at right angles to the surface. This gives the simplest analysis and also represents the worst case. The behaviour of the plate depends on its ductility, hardness and thickness and the size and shape of the projectile as well as on the impact speed. Definitions of plate thickness are very approximate and can vary. As the projectile gets blunter and the plate gets

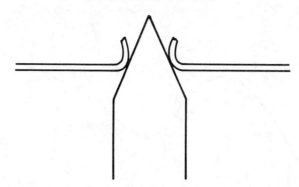

Figure 8.14 Perforation of thin plate by sharp-nosed projectile.

harder the transition from thin plate behaviour to intermediate plate behaviour occurs with thinner plates. In broad terms, thin plates are appreciably thinner than the projectile diameter; intermediate plates are approximately this diameter; and thick plates are appreciably thicker than this, merging into plates which are infinitely thick.

Except at very high velocities, thin plates of ductile material deform by bending. The plate may dish, bending overall, and there is severe local bending in the area close to the projectile. A small sharp-nosed non-deforming projectile is likely to perforate the plate at a fairly low velocity. The displaced material bends outward from the plate as shown in Fig. 8.14. The material from the centre of the hole cannot stretch to the projectile diameter but tears at about four locations, uniformly spaced round the circumference of the hole, leaving roughly triangular petals projecting from the back of the plate. At the other extreme a large blunt deformable projectile is likely to produce behaviour broadly similar to Fig. 8.15, which shows a typical load–deflection curve for a motor car door panel struck at right angles by the front of another car.

The reduction in the velocity of a sharp-nosed non-deforming projectile as it perforates thin plate can be estimated approximately for low and high velocities by ignoring dynamic effects in one case and plate strength in the other. Figure 8.16 shows theoretical velocity reductions derived from the plastic strain energy absorbed as a sharp-nosed projectile is pushed through a thin plate statically, and from a momentum interchange between a projectile and a plate of zero strength. Empirical relationships, also shown in Fig. 8.16, demonstrate a gradual transition from one type of behaviour to the other as impact velocity increases (Goldsmith, 1960, Fig. 153).

Overall bending can be analysed for a thin plate struck by a rigid projectile by assuming a deflected shape for the plate and equating the plastic strain energy absorbed to the kinetic energy lost by the projectile. For plates struck in

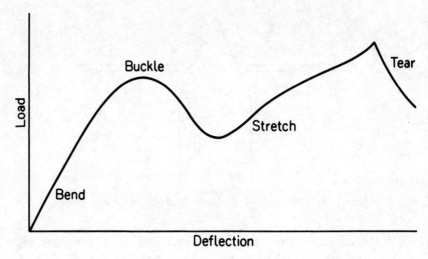

Figure 8.15 Penetration of a car door by the nose of another car.

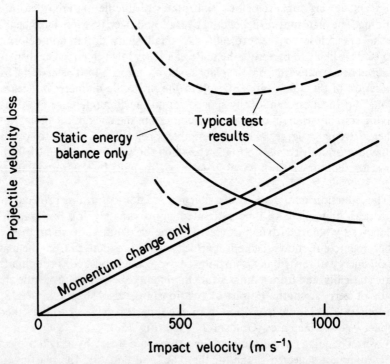

Figure 8.16 Loss of velocity when a sharp-nosed projectile perforates a thin plate.

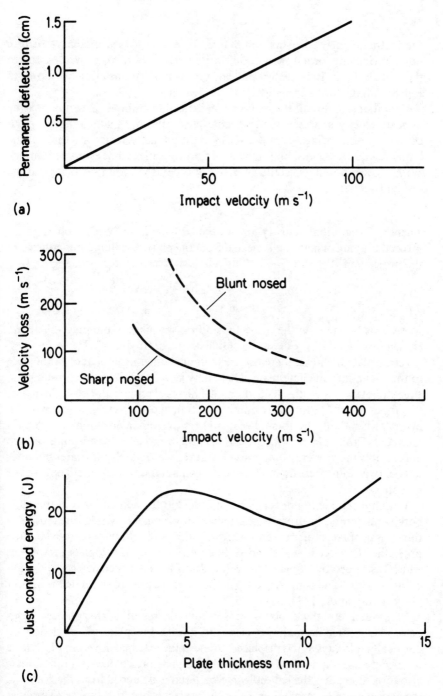

Figure 8.17 Behaviour of thin plates struck by rigid projectiles. (a) Permanent deflection of centre of a thin plate struck by a blunt rigid projectile. (b) Velocity loss of a rigid projectile perforating a thin plate. (c) Variation in just contained energy with plate thickness.

the centre the theoretical maximum overall deflection, measured at the centre but ignoring any local deformation, varies linearly with impact velocity up to the ballistic limit. Empirical results also show a linear variation, as shown in Fig. 8.17(a) (Calder and Goldsmith, 1971, Fig. 21).

Definitions of the ballistic limit vary in detail, but a simple one is that it is the impact velocity at which the projectile penetrates the plate completely but comes to rest in doing so. On this basis, if the impact velocity and the ballistic limit are known, the residual velocity of the projectile can be estimated simply. With a sharp-nosed projectile no material is detached from the plate and the residual velocity is

$$v_r = \sqrt{(v^2 - v_B^2)}$$

where v is the initial velocity and v_B the ballistic limit. With a blunt-nosed projectile a plug of material is ejected from the plate. Assuming this adheres to the projectile,

$$v_r = \sqrt{\left[\frac{m}{m + m_P}(v^2 - v_B^2) \right]}$$

where m is the mass of the projectile and m_P the mass of the plug. These are shown in Fig. 8.17(b) (Calder and Goldsmith, 1971, Figs 22, 23).

As the plate thickness increases, behaviour changes from that of a thin plate to that of an intermediate plate. There may be a transition zone, as shown in Fig. 8.17(c), in which the ballistic limit falls as the plate thickness increases (Onabe et al., 1984). In this region it would be more effective to use two thin plates rather than one thicker plate. There are other occasions in which a number of thin plates which bend overall as well as perforating absorb appreciably more energy than a single thicker plate which perforates but does not bend overall. There do not seem to be general rules which could be used for guidance.

With hypervelocity impact the projectile and the contact area of the plate both disintegrate. In protecting satellites from impact by micrometeorites a thin bumper plate is often used as shown in Fig. 8.18. This is penetrated but the projectile is reduced to a cloud of fine particles, liquid droplets or vapour, which is stopped by a thicker plate. It is difficult to test such devices at the full design velocity, and complex computer models are used to simulate the impact (de Rouvray et al., 1984, Figs 7.25, 7.26).

In moderately thick plates of soft ductile metal a sharp-nosed non-deforming projectile makes a clean hole, with the plate material flowing plastically sideways as it is displaced. Some of the material flows out of the hole to form a raised lip around the hole edge. If penetration occurs a raised lip is also formed around the exit hole. No material is detached from the plate. As the plate material becomes harder, or the projectile becomes blunter, a moving plug of material is formed ahead of the projectile nose. This ranges from about

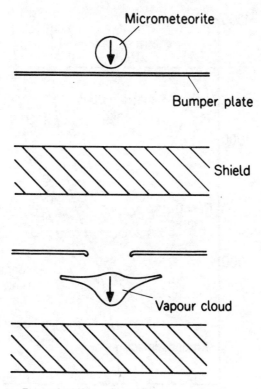

Figure 8.18 Use of a thin bumper plate.

one-third of the projectile diameter to the full diameter. If perforation occurs, this plug is ejected from the plate.

The overall effect of the angle of impact for a typical plate, equal in thickness to the projectile diameter, is shown in Fig. 8.19. Similar types of diagram can be produced to illustrate the effects of other parameters. These are usually derived empirically, apply to a limited number of materials, velocities and projectile shapes, and can become very elaborate (Zukas *et al.*, 1982, p. 177).

Bumper plates can be used to increase the effectiveness of plates of moderate thickness. The material, instead of being in a single plate, is divided into a relatively thin plate in front of a thicker one. If the projectile is brittle, with a tensile strength appreciably lower than its compressive strength, the compressive stress waves initiated by contact with the bumper plate reflect within the projectile as tensile stress waves and fracture it before it hits the main plate. The proportions and spacing of the plates need to be chosen with care.

As the impact velocity is increased beyond, say, $1000\,\mathrm{m\,s^{-1}}$ a critical velocity is reached at which a projectile of ductile material starts to flow like a

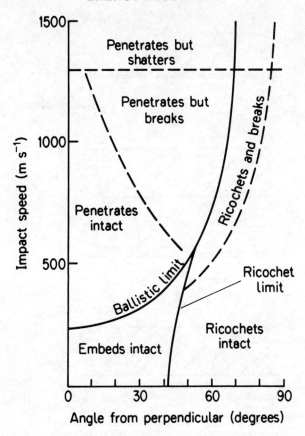

Figure 8.19 Projectile target phase diagram.

fluid even though its strength is still important. The projectile nose spreads and turns inside out forming a tube through which the rest of the projectile passes. This is shown in Fig. 8.20 (Zukas *et al.*, 1982, p. 166). The target material is forced aside in much the same way as by a non-deforming projectile, but the resulting hole is of appreciably larger diameter than the original projectile. The hole diameter remains the projectile diameter up to a critical velocity and then increases linearly with projectile velocity. In some tests the hole diameter had reached twice the projectile diameter at twice the critical velocity (Johnson, 1972).

If impact velocity increases still further the forces involved are so large that the material strength is unimportant and behaviour is effectively hydrodynamic. Standard hydrodynamic analysis can be used, and such impacts are easier to analyse than any other impacts except elastic ones. As a result there is an extensive literature. Impacts in this range are known as *hypervelocity* impacts.

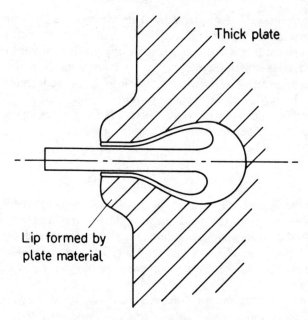

Figure 8.20 Solid projectile inverting at very high velocity.

The impact speed at which hypervelocity behaviour begins depends on the materials of the target and projectile, and it is not possible to specify it in absolute terms. It is necessary to define the onset of hypervelocity, and a number of definitions are possible. A common one is the complete pulverization of both projectile and target in the immediate vicinity of the initial contact point. On this definition, an impact velocity of about $5000\,\mathrm{m\,s^{-1}}$ is needed to produce hypervelocity behaviour of typical structural materials such as steel, aluminium and stone. A wide variety of behaviour is possible, especially when one of the impacting bodies has an appreciably lower hypervelocity speed than the other.

Metal targets of greater than intermediate thickness are unlikely to be perforated by non-deforming projectiles, but they can be penetrated by long rods impacting at very high velocity so that hypervelocity behaviour occurs at the rod nose. An appreciable length of the rod is likely to be consumed before the remaining length passes through the plate.

(d) Frangible targets

Concrete
Concrete is widely used both as a structural material and as a protection against projectiles. Structural concrete can withstand a substantial amount of penetration without collapsing, but the damage may be serious and it is often

desirable to protect the main structure. This can be done by compacted soil, sandbags or a layer of special protective concrete. When concrete is used as a protection against projectiles, it needs to be appreciably thicker than a metal plate, for equivalent performance, because it exerts a much smaller retarding force on the projectile. The resisting force of normal mass concrete may not be much more than 5% of that of steel armour.

Empirical formulae are widely used to specify the penetration of shells and bombs into mass concrete. A typical simple formula is

$$P = \frac{W}{D^2} K(v)^n$$

where P is the depth of penetration, W and D are the projectile weight and diameter, K is an empirical constant, v is the velocity and n is a number between 1 and 2. More elaborate formulae are in use but the large variability found in behaviour, and the fact that any empirical formula applies only to a limited range of conditions, raise the question of how far it is worth refining such formulae.

Usually v is specified in terms of a reference velocity at which some given behaviour occurs, and K is proportional to $1/\sqrt{\sigma_c}$, where σ_c is the compressive strength of the concrete. There is also an interaction between the size of the aggregate used in the concrete and the diameter of the projectile, so that K is not quite constant but varies slightly with D. Penetration is greater close to the free edge of a concrete slab, typically being 40% greater at eight projectile diameters from the edge.

At ballistic velocities, say greater than $150\,\mathrm{m\,s}^{-1}$, a projectile produces a crater on the front face of the slab. This is often known as *spalling* (see also Section 7.2.2(c)). The size of the crater increases rapidly with impact velocity up to about $500\,\mathrm{m\,s}^{-1}$ and then less rapidly. If a projectile perforates a slab, a crater is also formed on the back face. This is often referred to as *scabbing*. The scab is usually shallower and wider than the front crater and can inflict serious damage on items behind the slab. At velocities lower than ballistic, perforation can occur without a front crater. The two types of perforation are shown in Fig. 8.21. At ballistic velocities reinforcement limits the depth of the back crater to the distance between the back face and the nearest layer of reinforcement, but the area of the scab increases. Below ballistic velocities the back crater is less limited by the reinforcement as fragments of concrete can be pushed past it.

The impact velocity to just penetrate a concrete slab at velocities up to low ballistic velocities has been expressed empirically as

$$v_{\mathrm{crit}} = 1.3\sigma^{1/2}\rho^{1/6}\left(\frac{Dt^2}{W}\right)^{2/3} r^{0.27}$$

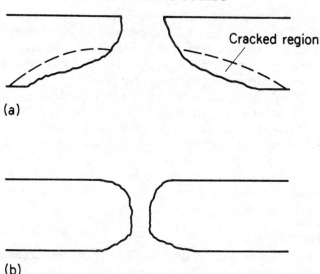

(a)

(b)

Figure 8.21 Perforation of concrete slabs. (a) Concrete perforated below ballistic velocities. (b) Concrete perforated at ballistic velocities.

where v_{crit} is expressed in metres per second, σ is the ultimate compressive strength of the concrete (Pa), ρ is the concrete density (kg m^{-3}), D is the missile diameter (m), t is the concrete thickness (m), W is the missile weight (kg) and r is the percentage of steel bending reinforcement. This is typical of the type of empirical formula in use, and there appears to be no straightforward way of comparing different formulae.

A lightweight concrete layer can be employed to reduce the peak loads on a concrete structure from heavy non-deforming projectiles at velocities up to, say, 50 m s^{-1}. Such a layer might be about 50 mm thick and of a concrete only about 10% of the strength of normal concrete, but it can reduce the peak impact loads in the underlying structural concrete by perhaps 70%. Performance depends on the shape of the projectile nose, but there seem to be no established general patterns of behaviour and development is largely by trial and error.

A fibre-reinforced concrete layer can greatly reduce damage caused by rifle bullets. Again there seem to be few established general patterns of behaviour and development is by trial and error. A layer about 125 mm thick and reinforced by steel fibres 25 mm long by 0.25 mm diameter, incorporated at random in the concrete up to a weight of 10% of the concrete weight, appears to provide significant protection.

It is also possible to use a layer of gravel set in a rubber-like matrix. The larger and harder is the gravel, the better is the penetration resistance. Gravel of approximately 25 mm to 35 mm diameter to about 60% of the total weight in a layer approximately 150 mm thick appears to provide almost complete protection from rifle bullets. With both the fibre-reinforced and the rubber-like protection layers, a major problem appears to be finding the cheapest and most easily applied materials (Anderson *et al.*, 1983).

Computer simulations have been made of aircraft crashing into reinforced containment vessels surrounding nuclear reactors. The results appear to be realistic but it is not feasible to check them at full scale. The impact speeds are about 200 to 300 m s^{-1}; the aircraft is destroyed apart from the engines, which are effectively solid and have to be stopped. A two-layer concrete protective screen has been proposed. The first layer destroys the aircraft and slows the engines before it is perforated. The second layer stops the resulting debris and brings the engines to rest.

Soil and sand
Compacted soil and sand can stop projectiles very effectively. The retarding force is usually assumed to be of the general form

$$F = A(a + bv + cv^2)$$

where A is the cross-sectional area, v is the instantaneous velocity and a, b and c are empirical constants. At impact velocities greater than about 60 m s^{-1} the v^2 term is dominant and the v term is generally omitted. The resulting simpler formula

$$F = A(a + cv^2)$$

leads to a depth of penetration of the general form

$$P = K_1 \log_e(1 + K_2 v_0^2)$$

where v_0 is the initial impact velocity and K_1 and K_2 are empirical constants.

The parameter of most general interest is the depth of penetration and, as with all impact analysis, it is extremely difficult to verify the details of behaviour from the final depth of penetration. As a result empirical formulae have been established for depth of penetration which sometimes have a logarithmic term in v^2 and sometimes do not.

There is considerable scatter in results and, even with nominally identical conditions, one projectile can penetrate twice as far as another. This has led to attempts to produce simple formulae of wide application. Figure 8.22 shows approximate average values for the depth of penetration of steel spheres into sand and rock. The curves were derived from results for a variety of shapes of projectile using the concept of *calibre density* or ballistic density. This uses the parameter

$$\rho_c = \frac{W}{D^3}$$

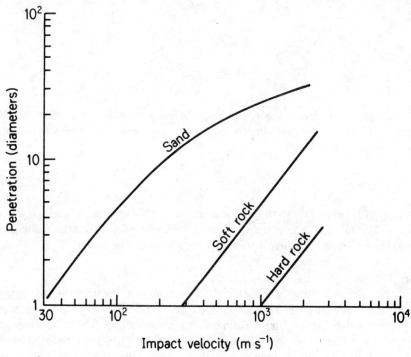

Figure 8.22 Very approximate guide to penetration by steel spheres.

where ρ_c is the calibre density, W the projectile weight and D the projectile diameter. Note that here, and in the subsequent discussion, density is defined as weight per unit volume, not mass per unit volume (Kornhauser, 1967, Fig. 3.3).

The assumption is made that, for a given impact velocity, depth of penetration P varies linearly with calibre density, so that

$$\frac{P_s}{P} = \frac{\rho_{cs}}{\rho_c}$$

where P_s is the penetration and ρ_{cs} the calibre density of a steel sphere. The latter is given by

$$\rho_{cs} = \rho_s \left(\frac{\pi}{6} D^3 \right) \frac{1}{D^3}$$

$$= \frac{\pi \rho_s}{6}$$

where ρ_s is the density of steel. From this,

$$P_s = P\frac{\pi\rho_s}{6}\frac{1}{\rho_c}$$

$$= P\frac{\pi\rho_s}{6}\frac{D^3}{W}$$

Reversing the process, the penetration of projectiles of other shapes and materials is found from the curves of Fig. 8.22 by multiplying the penetration for steel spheres by

$$\frac{W}{D^3}\frac{6}{\pi\rho_s}$$

A projectile penetrating soil sometimes rotates so that it is travelling sideways or backwards, and sometimes follows a J-shaped path so that it reaches a maximum depth and then moves part way back to the surface.

(e) Fabrics

Fibre-reinforced composite materials do not appear to be used as energy absorbers, but cloth woven from strong synthetic fibres is used to protect people and delicate machinery from damage by projectiles. The basic aim is to prevent penetration on the assumption that the resulting distributed load from the fabric is much less damaging than the original concentrated load from the projectile. The fabrics used are woven from nylon or aramid fibres which are sometimes supplemented by thin aluminium sheets. The aluminium appears to be included mainly to reduce costs.

The design of such fabrics appears to be based on trial and error, and they are tested with both large and small projectiles representing the range of bullets likely to be fired at them. A small projectile has lower kinetic energy but it also impacts a smaller number of fibres. Performance is assessed by comparative tests, the main criterion being the velocity or energy needed to just penetrate the fabric (Miner, 1984).

9

Structures

9.1 INTRODUCTION

9.1.1 General

(a) Complexity of analysis

When deflections are small and materials are linearly elastic, general relationships can be established among loads, deflections and stresses in structures. When deflections are large, materials are non-linear and time-dependent effects occur, linear small deflection behaviour is still often used as a basic for analysis. Often there is not enough information available to make a more comprehensive analysis worth while. Again, there may be considerable scatter in behaviour so that a simple analysis incorporating statistical variations may be better than a more elaborate deterministic one. For static loading and steady-state vibrations, well-established semi-empirical rules exist relating actual behaviour to simple linear analysis. These may not apply to impact loading.

In low and medium velocity impacts, say up to $60 \, \mathrm{m \, s^{-1}}$, the overall behaviour of the structure is of interest and, unless the load is very concentrated, detailed analysis of conditions at the point of impact is seldom needed. At high velocities, say above $150 \, \mathrm{m \, s^{-1}}$, overall structural behaviour can usually be ignored because, except for blast loading of the structure as a whole, deformation is concentrated at the impact point. In between, both types of deformation may be important. This chapter is confined to overall structural response.

It is possible to analyse a structure at varying degrees of complextiy. In comparing results from different analyses, only equivalent parameters can be used. For instance, some aspects of the overall response of a complex structure to a given force–time pulse can be reproduced by very simple models which give the correct motion of a given point on the structure. It is not possible to find the stresses in the structure directly from such a model; they need to be found indirectly by studying the effects of this motion on the actual structure.

Simple models can be used to optimize the structure overall and to give insights into the main patterns of behaviour. More complex models can then be used to investigate behaviour in detail either on the structure overall or in areas of particular interest. Usually simple models can be derived from first principles but, with more complex models, it is better to use existing standard computer programs or to employ specialist analysis firms, because considerable experience and skill is needed in both modelling and interpretation of results.

(b) Plastic hinges

The concept of the plastic hinge is widely used in the analysis of structures which bend plastically. Such bending is often concentrated in a small area, and the simplest form of analysis treats this as equivalent to a hinge of zero length which requires a constant moment to turn it. More sophisticated analyses extend this to cover a plastically deformed area of finite length and a varying turning moment.

Sometimes a plastic hinge is stationary and remains in its original location as it rotates. At other times the hinge travels along the deforming member. In this case there is first bending and then straightening of the plastically deformed materials as the hinge passes, and much more energy is dissipated than would appear from the final deformed shape.

Travelling plastic hinges are found in thin walled structures which fail by local buckling. As deformation proceeds the buckles change shape, and this involves the hinges changing their location. In the static deformation of solid structural members, plastic hinges are stationary, but travelling plastic hinges have been postulated to explain the final deformed shapes of solid structural members loaded by impact. It is difficult to reconstitute behaviour during an impact from final displacements and the widespread existence of such travelling hinges has not been proved. They have been detected in very small beams loaded by explosives, but there is considerable doubt as to whether they exist with larger structural members and less rapid loading (Florence and Firth, 1965).

9.1.2 Basic relationships

(a) Scale effects

Much of the background experimental work on the impact loading of structures has been on small specimens with maximum overall dimensions less than one metre. With fairly small thin walled structures, up to about the size of motor cars, testing is usually done at full size. With larger thin walled structures, such as aircraft, a small amount of full scale testing of complete structures has been reported and components are often tested at full size. With very large structures, such as buildings, bridges and ships, full scale testing is almost unknown and testing is done on scale models.

There are dangers in extrapolating results from a smaller to a larger scale even when using the appropriate dimensional analysis techniques. There are three reasons for this, two of which apply to scaling generally and one of which is specific to impact behaviour. These are that it is not possible to scale all parameters correctly, and so a choice has to be made of key parameters; that it is not possible to construct a completely accurate model; and that there can be significant anomalies in impact behaviour which are not fully understood.

Present knowledge indicates that, as long as deformations remain ductile, results from scale models can be used, although allowance needs to be made for time-dependent effects in both the material properties and the dynamics of the deformations. Where tearing, cutting and brittle fracture occur, results from scale models cannot be relied on and the full scale structure is appreciably weaker than the model tests would predict. Typically full scale deflections may be twice those predicted by extrapolation from a quarter-scale model. A complicating feature is that these non-ductile patterns of deformation become more likely as the size of a structure increases, so that a small scale model is likely to deform in a ductile manner whether the full size structure does or not (Jones, 1984).

(b) Dimensionless parameter groups

Even with a simple beam impacted by a simple rigid mass, there are six independent variables – beam length, mass and stiffness, the impacting mass and its velocity, and the time at which behaviour is being considered. There are also two dependent variables – the location and magnitude of the maximum deflection. In any specific case, five of these are known in advance, and analysis can be carried out using only the two dependent variables and time.

For general analysis it is necessary to consider the effects of all eight variables and, to make this tractable, they are combined into groups which are usually chosen to be dimensionless. These groups are usually selected for ease of analysis rather than for physical insight and, though they come to have a tangible meaning for people who use them regularly, they are confusing to the non-specialist. To date no common usage has emerged analogous to the use of Reynolds number in fluid dynamics, and different authors use different groups when analysing similar problems. Even if they use the same groups they may express them differently or use different symbols. In this chapter results are presented, as far as possible, in terms of simple physical parameters. Some representative parameter groups are given in Table 9.1. It should be remembered that a parameter discussed in this chapter in simple general terms is likely to be expressed in the literature in a non-dimensional way and may appear incomprehensible at first sight; various constants will be incorporated and there will be some variation in the symbols used, so all published results need to be checked individually.

Table 9.1 Dimensionless parameter groups.

Purpose of parameter	Symbol	Parameter group
Mass ratio	K	M/mL
Beam length	S	$(m/M)L = 1/K$
Distance to plastic hinge	ξ	x/L or $(m/M)l$
Time	η	vt/L
Kinetic energy	α	$\dfrac{T}{M_0} = \dfrac{Mv^2}{2M_0}$ or $\dfrac{mLv^2}{2M_0}$
Duration of impact	$\eta/\alpha K$	$\dfrac{vt}{L}\dfrac{2M_0}{Mv^2}\dfrac{mL}{M} = \dfrac{2mM_0t}{m^2v}$
Rate of energy absorption	α/η	$= \dfrac{mLv^2}{2M_0}\dfrac{L}{vt} = \dfrac{mL^2v}{2M_0t}$
Plastic hinge rotation	θ/α	$2M_0\theta/mLv^2$
Axial tensile force	β	N_0L/M_0 For beams of geometrically similar cross-section, $\beta \propto L/h$
Kinetic energy when axial tensile force is present	$\zeta = \sqrt{(\alpha\beta)}$	$\sqrt{\left(\dfrac{mLv^2}{2M_0}\dfrac{N_0L}{M_0}\right)} = \dfrac{Lv}{M_0}\sqrt{(mN_0)}$ For beams of geometrically similar cross-section, $\zeta \propto \dfrac{Lv}{\eta}\sqrt{\left(\dfrac{m}{N_0}\right)}$

9.2 STRUCTURAL COMPONENTS

9.2.1 Beams

(a) Solid

In solid or thick walled beams of ductile material the cross-section is assumed to retain substantially its original size and shape even though considerable plastic flow may occur parallel to the beam's longitudinal axis. To simplify the present discussion, it is confined to beams of uniform cross-section along their length so that the stresses produced by bending vary directly with the bending moment. The main discussion deals with the *cantilever*, which is a beam rigidly held at one end and unsupported at the other. This demonstrates behaviour in its simplest form.

If a static force F is applied to the free end of a cantilever at right angles to the longitudinal axis, as shown in Fig. 9.1, it produces a transverse shear force

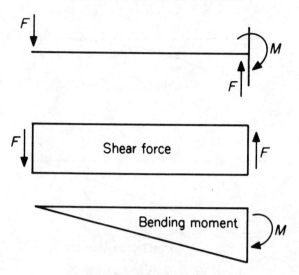

Figure 9.1 Cantilever with load at tip.

which remains constant along the beam and is reacted by a force $-F$ at the fixed end. It also produces a varying bending moment $M = Fl$, where l is the distance along the beam from the free end. In most beams the stresses produced by the maximum bending moment are much larger than the transverse shear stresses, but in beams which are very short, or vibrating in a mode with a short wavelength, the two sets of stresses are comparable.

Under a static load the beam bends in one of two basic ways. If the stress remains below the yield stress the beam bends into a smooth continuous curve but, once plastic flow occurs, deformation tends to concentrate in a single plastic hinge about which the beam rotates. These are shown in Fig. 9.2. The elastic deformation can be expressed by the curvature $K = \mathrm{d}\theta/\mathrm{d}x$, where θ is the angle between the deflected and undeflected longitudinal axis and x is the distance along the beam, or by the radius of curvature r, which is the radius of the circle which has the same curvature as the beam at any point. These are inversely proportional, a large curvature giving a small radius of curvature. Simple elastic bending theory gives

$$r = \frac{EI}{M} \tag{9.1}$$

where E is Young's modulus for the material and I is the moment of inertia of the cross-section. For a beam of constant cross-section, EI is constant and

$$r \propto \frac{1}{M} \tag{9.2}$$

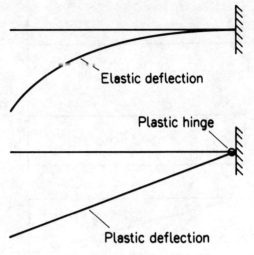

Figure 9.2 Deflection of a cantilever.

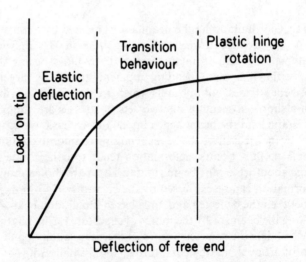

Figure 9.3 Load–deflection curve for a cantilever.

At an idealized plastic hinge r is not specified but M is constant regardless of the hinge rotation, and the beam deformation can be expressed in terms of the angle through which the hinge rotates. Both types of behaviour can be compared if the deflection δ of the free end of the cantilever is plotted against F, as shown in Fig. 9.3. There are three stages: an elastic stage in which δ varies linearly with F; an elastic–plastic transition; and a plastic stage in which F

remains substantially constant (Norris *et al.*, 1959, Fig. 1.7). If significant rotation occurs about the plastic hinge the effective length of the beam decreases, and so *F* needs to increase to maintain bending moment. In simple analysis this is usually ignored.

If the energy absorbed in plastic bending is appreciably greater than the maximum energy the beam could store in elastic bending, elastic behaviour can be ignored. The maximum strain energy the beam can store elastically occurs when the whole beam is subjected to a constant bending moment M_y which just produces the yield stress. This energy can be shown to be

$$V_{max} = \frac{M_y L}{2EI} \tag{9.3}$$

where *L* is the total length of the beam. It is twice the energy stored elastically in the cantilever when M_y is reached at the fixed end because the mean bending moment in this case is only half the maximum bending moment, as can be seen in Fig. 9.1.

If the ratio

$$R = \frac{\text{plastic bending energy}}{V_{max}}$$

is about 3, plastic hinge analysis gives a first approximation to impact behaviour but elastic vibrations can have a noticeable effect. If $R > 10$, elastic behaviour can be ignored and plastic hinge analysis is accurate enough for most purposes (Bodner and Symonds, 1972).

The curve in Fig. 9.3 is similar to the stress–strain curve for a ductile material and similar approximations can be used. Elastic and plastic behaviour are easier to analyse than transition behaviour and simple analysis concentrates on them.

The maximum stresses produced by bending are tensile on the convex side of the beam and compressive on the concave side and they act parallel to the longitudinal axis. In simple elastic bending theory the bending stress occuring anywhere in the cross-section is

$$\sigma = \frac{My}{I} \tag{9.4}$$

where *y* is the distance from the neutral axis which is located at the centroid of the cross-section. Combining Equations 9.2 and 9.4 for a beam of constant cross-section gives

$$\sigma \propto M \propto \frac{1}{r} \tag{9.5}$$

If the static force is replaced by a sinusoidally varying force $F_1 \sin \omega t$, the beam vibrates transversely in a number of modes each of which has its own natural

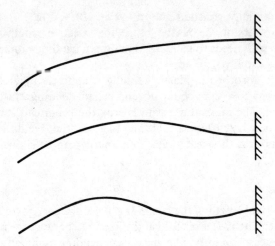

Figure 9.4 First three mode shapes for a vibrating cantilever.

frequency and its own characteristic shape. The first three mode shapes for a cantilever are shown in Fig. 9.4. The natural frequencies and characteristic shapes can be found directly by forming and solving the equations of motion, but analysis can become complicated and often approximate methods are used. Generally these are based on energy methods using simple, assumed mode shapes (Hurty and Rubinstein, 1964, Chapter 4).

Axial vibrations of the beam also occur whilst it is vibrating transversely. These may need to be considered if propagation of bending along a beam is being analysed in detail, but usually they are ignored. They are sometimes called *symmetrical vibrations* and the transverse vibrations, in this context, are called anti-symmetrical. This can be confusing because the transverse vibrations have both symmetrical and anti-symmetrical modes.

For a given maximum value of the sinusoidal force the maximum transverse displacement varies with $1/\omega^2$, as discussed in Chapter 6. If a number of modes are present simultaneously, and none of them is being excited preferentially, it is usually assumed that the maximum transverse displacement in each mode is inversely proportional to the square of its natural frequency. The maximum displacement of any point on the beam is then given accurately by only the fundamental mode, or at most the sum of the first few modes.

The behaviour of the maximum elastic bending stress is more complicated because two opposing factors are present: the smaller transverse displacement gives a lower bending stress for a given wavelength, but the shorter wavelength of a higher mode gives a smaller value of r for a given transverse displacement, leading to a higher bending stress. The two effects appear to be comparable in size, leading to ill-conditioned equations, and a large number of modes may need to be included before the maximum elastic bending stress converges to a consistent value (Goldsmith, 1960, Fig. 57). It seems likely that, in such a case, the onset of yield is determined more by imperfections in the beam or variations in the loading than by the effects of very high harmonics. This appears to be a situation in which a statistical approach based on a simple analysis would be more appropriate than a detailed deterministic analysis.

When the tip of a cantilever is impacted transversely by a mass there is an interaction between the beam and the mass which can be dealt with generally as discussed in Section 6.3. If only the first mode needs to be considered the beam can be regarded as simple spring–mass system. Higher modes can be dealt with using steady-state beam vibration theory and Fourier analysis.

In simple theory the interaction between the beam and the mass is often ignored and all the kinetic energy of the mass is assumed to be transferred to the beam. This is equivalent to considering the mass as initially stationary and attached to the free end of the cantilever, and then subjected to an impulse which causes an instantaneous transverse velocity. The analysis discussed here assumes that the mass remains attached to the beam and that there is no dissipation of energy in the region of contact between the mass and the beam. Elastic waves propagates as shown in Fig. 9.5. An initial transverse shear stress propagates along the beam at the shear stress wave velocity C_T, and the total shear stress builds up by successive reflections between the fixed end and the impacting mass. A bending wave also propagates along the beam at velocity C_b, which can be derived in a similar way to other stress wave velocities (Kolsky, 1963, pp. 48–53, 68–73). C_b is not constant but varies with wavelength, short wavelengths propagating more quickly than long wavelengths. In the simplest theory, which includes only bending stresses and transverse intertia, very short wavelengths propagate instantaneously. At these short wavelengths transverse shear stress and rotational inertia are important and, if

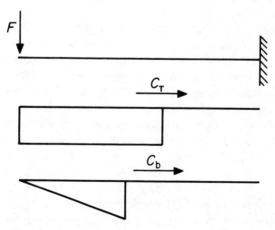

Figure 9.5 Propagation of shear and bending waves along a cantilever.

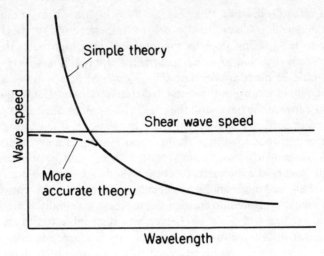

Figure 9.6 Propagation of bending waves.

these are included, short wavelengths propagate at a velocity slightly slower than C_T. The variation in C_b with wavelength is shown in Fig. 9.6.

When the deformation of the beam is predominantly plastic $(R > 3)$, travelling plastic hinge theory has been used. In this theory a plastic hinge forms at the free end of the beam and travels towards the fixed end at velocity C_H, as shown in Fig. 9.7(a). The hinge propagates in a similar way to the elastic wave front of the fundamental bending mode. Ahead of it the beam is stationary, whilst behind it the beam rotates as a rigid body. Any kinetic energy remaining in the beam and the impacting mass when the hinge has reached the fixed end is absorbed in further rotation of the hinge at this location.

The travelling plastic hinge deforms the beam into a curved shape. The relative amount of energy absorbed by the hinge as it travels and in its final location varies with mL/M, where M is the moving mass and mL is the mass of the cantilever. It is convenient to use a dimensionless beam length $S = (m/M)L$ and a dimensionless energy input $\alpha = T/M_0$, where T is the initial kinetic energy of the moving mass and M_0 is the moment needed to rotate the plastic hinge. On this basis, plots are shown of the energy absorbed (Fig. 9.7(d)) (Bodner and Symonds, 1972, Fig. 4) and the deflected shape (Fig. 9.7(d)), together with some test results on small solid cantilevers of steel and aluminium (Fig. 9.7(c)) (Ting, 1965, Fig. 7). Note that though this simple theory agrees with the final deformed shape (Ting, 1965, Fig. 2) it is open to doubt whether such a travelling plastic hinge exists.

If a transverse velocity is applied to the fixed end of the cantilever, elastic behaviour is similar except that the elastic waves travel along the beam in the

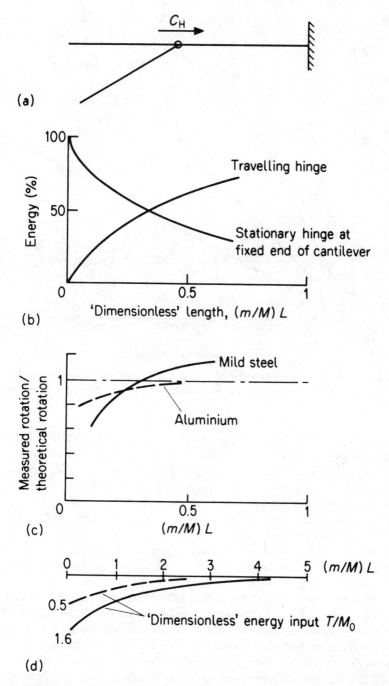

Figure 9.7 Travelling hinge analysis of small cantilever. (a) Travelling plastic hinge.
(b) Energy absorbed by plastic hinges. (c) Overall rotation of free end of cantilever.
(d) Dimensionless deflected shapes.

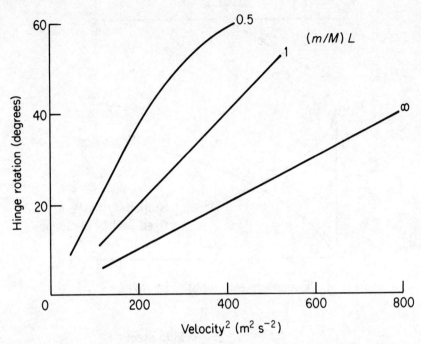

Figure 9.8 Rotation of hinge at fixed end of cantilever subjected to a sudden velocity at its fixed end (in this case, M is a mass attached to the free end of the cantilever).

opposite direction. Plastic hinge behaviour is different, however, with the travelling plastic hinge rotating in the opposite direction to the one at the fixed end of the cantilever and contributing very little to the overall deformation. As most of the deformation occurs at the hinge at the fixed end, it is convenient to give deformation in terms of the rotation of this hinge (as shown in Fig. 9.8) and the travelling hinge can generally be ignored (Mentel, 1958, Fig. 12).

If the transverse impact force is uniformly distributed along all or part of the length of the cantilever the behaviour depends on a number of variables, and there appear to be five possible modes of deformation (Johnson, 1972, pp. 278–9).

The behaviour of other types of beam can be treated in much the same way as the cantilever but the effects of the end supports during an impact can be very complicated. A beam with simply supported ends and a static transverse force applied at mid span is shown in Fig. 9.9. The ends are held in a fixed position relative to the initial longitudinal axis of the beam but are free to rotate. If they are also free to move towards each other as the beam bends, behaviour is the same as that of two cantilevers with a common fixed end at the mid span of the beam. This is shown in Fig. 9.9. If the ends are restrained

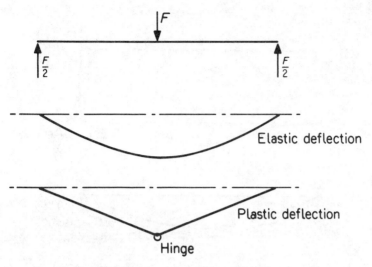

Figure 9.9 Deflection of a simply supported beam.

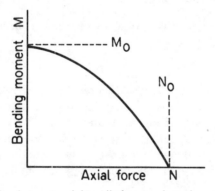

Figure 9.10 Interaction between axial tensile force and maximum bending moment in the fully plastic state.

axially, an axial tensile force occurs which increases as the beam deflection increases. This does not have a large effect on elastic behaviour unless the beam is so slender that it acts more as a string than a beam, but it has an appreciable effect on plastic behaviour. There is an interrelation between the plastic bending moment and the axial tensile force, of the type shown in Fig. 9.10 (Symonds and Mentel, 1958, Fig. 2(b)), which arises because the total resultant plastic stress cannot exceed a given maximum value. It is necessary to use a multidimensional flow rule of the type discussed in Chapter 3; the exact

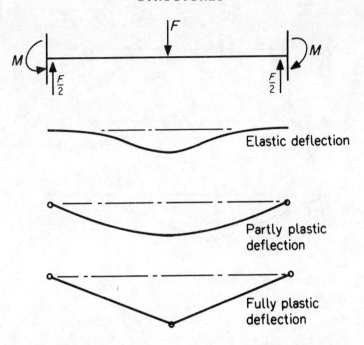

Figure 9.11 Deflection of a beam with built-in ends.

relationship depends on the flow rule chosen, but in all cases the plastic hinge moment decreases as the beam deflection increases. Detailed analysis can become very complicated (Samuelides and Frieze, 1983).

A beam with built-in ends cannot rotate at the end supports. If it is loaded statically it deforms as shown in Fig. 9.11. As the load increases, a stationary plastic hinge forms at each end and behaviour is then similar to that of a simply supported beam.

If a central transverse force arises from an impact, behaviour depends on the length of the beam relative to its depth and relative to the speed of sound in the material as well as on the type of end support. Generally it appears that, because of the complexity of the equations and the number of variables, any specific case is worked out numerically. Two extreme conditions shown in Fig. 9.12 have been worked out in general terms – a short beam with no supports, and an 'infinite' beam which is long enough for deformation to be complete before a stress wave can return from the end supports (Johnson, 1972, Fig. 6.3). The first is the same as two cantilevers displaced at a common fixed end at the centre of the beam. The second has been analysed using elastic bending waves or travelling plastic hinges (Johnson, 1972, Fig. 6.2).

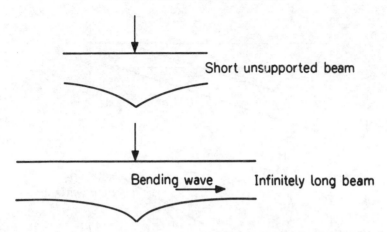

Figure 9.12 Beams in which end supports do not influence impact bending.

(b) Thin walled

Under static loading, thin walled beams behave initially in much the same way as thick walled ones, bending overall into a continuous smooth curve. A second stage of behaviour begins with local buckling on the concave side of the beam where the bending stresses are compressive. If the beam is relatively thick walled a single plastic buckle forms and the force–deflection curve for a cantilever is as shown in Fig. 9.13. The force falls as the buckle forms but then the two ends of the buckle meet and the force rises again. If the beam is very thin walled the local buckling is initially elastic and is likely to consist of a number of buckles, one or more of which then become plastic. The force does not rise again but keeps falling continuously in much the same way as the axial force on a simple strut. This is also shown for a cantilever in Fig. 9.13. As the wall thickness changes the pattern of behaviour changes progressively from one type to the other.

When the local buckles have deformed enough for plastic hinges to occur, plastic analysis can be used. As with the plastic bending of solid beams, transition behaviour is complicated and simple analysis is usually restricted to either linearly elastic or rigid plastic behaviour. A simple system of plastic hinges for the buckling of a thin walled beam is shown in Fig. 9.14. These hinge systems can become quite complicated, but for static loading they have been systematized into a number of basic patterns and their behaviour analysed (Murray, 1983). Under dynamic loading buckling can occur in higher modes with shorter wavelength buckles. Within the buckle patterns, travelling plastic hinges occur because the individual buckles need to change shape in order to keep folding. The plastically deformed areas do not travel bodily along the beam, and so the plastic deformation is concentrated in the initial area.

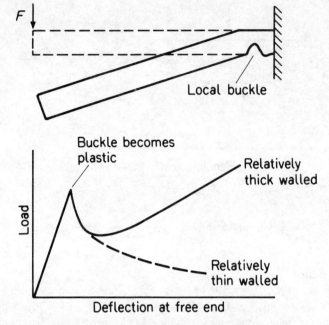

Figure 9.13 Bending of thin walled cantilever.

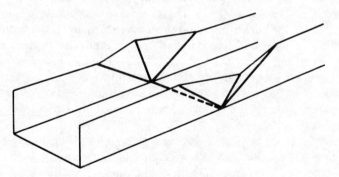

Figure 9.14 Plastic hinges in a typical local buckle.

9.2.2 Struts

(a) Solid

Struts are straight, relatively slender structural components which are compressed axially along their length. If a strut is compressed by a static axial force F it remains straight initially and the only deflection is due to the compressive strain in the material. At a critical value of force F_c the strut

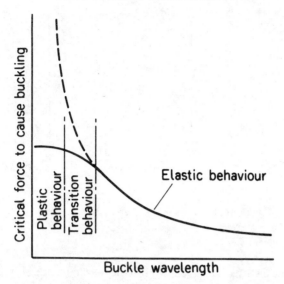

Figure 9.15 Variation in critical force with buckle wavelength for a simple strut.

becomes unstable and bends. The load which it can support falls and, if there is no other constraint on axial deflection, the strut collapses. If deflection is controlled the falling force–deflection curve can be observed.

Theoretically a strut can buckle into an infinite number of modes which have the same shape as the characteristic shapes of the transverse vibrations of a beam. The critical force F_c increases as the wavelength of the buckles decreases, as shown in Fig. 9.15, so that, unless some form of transverse restraint forces a shorter wavelength, buckling occurs at the longest wavelength and the lowest critical load. In dynamic loading the inertia of the strut mass resists transverse movement; this is greater for the longer wavelength buckles, and so shorter wavelength buckling is likely to occur with a correspondingly higher critical load.

Linearly elastic buckling only occurs with very slender struts, but elastic buckling behaviour forms a useful reference with which other types of behaviour can be compared. The theoretical elastic force–deflection curve rises to a sharp narrow peak, as shown in Fig. 9.16(a). A small disturbance anywhere near this peak is enough to move the strut from the straight to the buckled state and it cannot then return to the unbuckled state. As a result the elastic buckling load is very sensitive to small imperfections in the strut or the axiality or smoothness of loading, and there is a good deal of scatter in test results even with carefully prepared specimens. Working rules for practical struts loaded statically allow for this variability by incorporating semi-empirical factors based on assumed initial imperfections and assumed criteria for strut collapse.

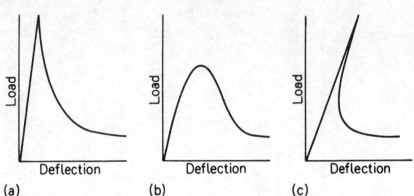

Figure 9.16 Load–deflection curves for different types of buckling. (a) Elastic. (b) Plastic. (c) Snap-through.

Most practical struts are not sufficiently slender to buckle elastically overall and they buckle either elastoplastically or plastically. Behaviour is broadly similar to elastic buckling but the transition from stable to unstable behaviour is gentler, as shown in Fig. 9.16(b), and there is less scatter in test results.

In a few cases of practical interest the theoretical load–deflection curve is re-entrant and the strut cannot follow this but jumps from the unbuckled to the buckled state. This is called snap-through buckling, and is shown in Fig. 9.16(c). The actual buckling load is likely to be appreciably less than the maximum theoretical value and subject to even greater scatter than simple slender struts.

In all cases, as deformation progresses after buckling, the effects of initial imperfections become less important and, once appreciable buckling is present, subsequent behaviour is consistent and predictable. As with beams, thick walled ductile struts bending elastically form a smooth curve but extensive plastic deformation tends to be concentrated at plastic hinges.

If the axial force is $F = F_0 + F_1 \sin \omega t$ so that it has both a constant and a sinusoidally varying component, vibrations occur. As with the transverse vibrations of a beam, both axial or symmetrical and transverse or anti-symmetrical vibrations can be present. The axial vibration follows the usual pattern but the transverse vibration behaves quite differently and is of a type called *parametric vibration*. Theoretically it can occur in a number of modes, but only the fundamental mode appears to have been observed experimentally. This has a frequency of $\omega/2$ and can develop into resonance over a range of different frequencies as the steady component F_0 varies (Evensen and Evan-Iwanowski, 1966). Once parametric resonance occurs the strut collapses.

Parametric vibrations are complicated and dynamic buckling modes can be analysed more simply by buckling theory. Consequently elastic impact buckling is not analysed in terms of steady-state parametric vibrations in the way that the elastic response of beams to transverse impact is analysed in terms of steady-state transverse vibrations.

Short wavelength buckles develop more quickly than long wavelength ones but initial imperfections also strongly influence the rate at which buckles develop. Usually it is assumed that small initial imperfections are present but are of such a nature that the wavelength effect is dominant but with some scatter in results. Large initial imperfections override the wavelength effect but it is not clear how large these need to be to cause this, nor how far a buckle needs to develop before it can be regarded as established irreversibly as dominant. If it is necessary to ensure that a given mode of buckling is dominant, a solid strut needs to be preformed in this mode.

If the buckle wavelength, the statistical distribution of initial imperfections and the amount of buckling needed to produce collapse are known, it is possible to calculate the dynamic axial load which a strut can withstand. The inertia of the strut resists the transverse acceleration and increases the end load needed to produce buckling. Results for a simple pin-ended strut buckling in the fundamental mode are shown in Fig. 9.17. The load duration is given in

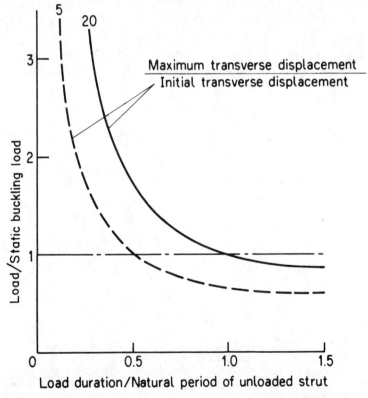

Figure 9.17 Response of initially slightly curved strut to half-sine compressive load pulse.

terms of the fundamental period of transverse vibration when no axial load is present (Davidson, 1953).

If the dynamic load is much greater than the static buckling load, higher modes of buckling occur. Short elastic buckles in small very thin struts develop in times comparable with the time it takes for an elastic wave of compressive stress to travel the buckle length. The stress wave front travels several wavelengths ahead of observable buckles developing and it is assumed, in analysis, that buckles develop under a constant axial force. This also appears to happen with a plastic stress wave producing plastic buckles (Abrahamson and Goodier, 1966).

With carefully conducted tests on very straight struts, buckles have been found to occur close to the theoretical preferred wavelengths with some scatter, as shown in Fig. 9.18 (Lindberg, 1965, Fig. 3). Under less controlled conditions the scatter is likely to be much greater and a simple first approximation to the buckle wavelength is probably sufficient. If transverse inertia is ignored and elastic behaviour is assumed, it can be shown very simply that the buckle wavelength is

$$\frac{\lambda}{k} = \pi \sqrt{\left(\frac{C}{v}\right)} \tag{9.6}$$

where k is the radius of gyration, C is the stress wave velocity and v is the impact velocity. If transverse inertia is included, the wavelength increases to

$$\frac{\lambda}{k} = \pi \sqrt{\left(\frac{2C}{v}\right)}$$

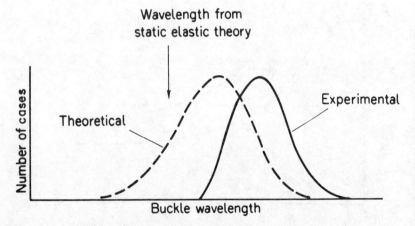

Figure 9.18 Predicted and measured buckle wavelengths in carefully conducted impact tests on thin aluminium strips.

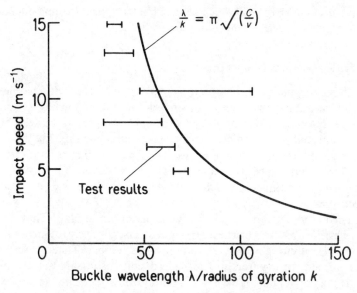

Figure 9.19 Wavelengths of small plastic buckles.

Some test results for permanent plastic buckles are shown in Fig. 9.19 (Johnson, 1972, Fig. 2.34). With a long strut, buckling may be confined to the impact end and the buckle wavelength may increase with distance from this end. If the reflected stress wave is compressive this may cause a similar pattern at the far end. Once plastic deformation has occurred the buckle pattern remains substantially unaltered, and travelling plastic hinges of the type postulated for beams loaded dynamically do not seem to occur with the simple buckling of thick walled struts.

(b) Thin walled

Under static axial load a thin walled strut buckles overall if it is sufficiently slender. Local buckling occurs on the side in which the bending stress is compressive and subsequent bending is generally similar to that of a thin walled beam. If the strut is too short to buckle overall it remains straight, but the walls buckle locally into a wavelength close to a preferred value which is determined by the properties of the material and the geometry of the cross-section. The value of F_c for these short struts is substantially independent of the length of the strut but varies slightly depending how close the length is to an exact multiple of the preferred local buckle wavelength.

The load–deflection curve for a thin walled strut crumpling entirely by local buckling is broadly similar to that for an axially loaded thin walled tube shown in Fig. 8.8(a). After an initial peak value at which buckling starts, the load fluctuates about a fairly steady mean value. Under impact loading buckling is likely to occur at modes higher than the static one.

9.2.3 Brittle structural components

Beams and struts made of brittle material may be reinforced with ductile rods as in reinforced concrete. Such members seldom fail completely as soon as fracture occurs; instead, they may rotate through an appreciable angle about a seriously damaged area in which the brittle material is shattered but the ductile reinforcing remains. Such behaviour is substantially similar to rotation about a plastic hinge and can be treated in the same way, although it may be much more difficult to estimate the moment needed to rotate the plastic hinge (Ammon, 1984).

Advanced composite materials are used mainly in aircraft. They usually consist of brittle fibres in a brittle matrix, but can be relatively tough because of inherent crack blunting properties and the large number of energy absorbing surfaces formed during fracture. Advanced composites are susceptible to barely visible impact damage when fairly minor low speed impacts such as a dropped spanner cause internal damage such as delamination within the composite. This can seriously weaken the structure, especially those members loaded in compression where the delaminated region can buckle at a load well below the original one (Cantwell, 1984).

9.3 STRUCTURAL MODELS

9.3.1 General

There are two main ways in which mathematical models of structures loaded by impact differ – the analysis method and the level of complexity. Analysis methods can be almost entirely mathematical, almost entirely empirical, or hybrid. They can range from very simple to very complex or can be simple in some areas and more complex in others.

Each category has its strengths and weaknesses. Experimental results are realistic but they may not be repeatable, and testing can be costly and time consuming. Results may be difficult to interpret; it may not be possible to assess what changes should be made; and the process can take so long that the results are known too late to influence the design. Mathematical models can be produced without experimental testing; the same model can be used in many simulations; and the effects of design changes can be assessed. They may be difficult to produce and may not be accurate, and important aspects of behaviour may not be foreseen. Hybrid models, whilst possessing the virtues of both approaches, may also suffer from the drawbacks of both.

Very simple models are easy to understand and give insights into patterns of behaviour, but it may be difficult to relate them to real structures. Initial optimization of a design can be done with simple models and more detailed models can be used for final design. Very detailed models can be difficult to understand and errors can go unnoticed. There may be problems with the stability and accuracy of complex computer programs; they can take a long time to develop and may use large amounts of computer time. The computing time can be reduced, at the expense of more complex programs, by incorporating detailed areas into a simpler model and by varying the length of each time step to just meet the minimum requirements of that step.

In *explicit analysis* the parameters of interest are expressed directly in terms of each other. In impact analysis, methods are called explicit if the displacements during a time step from t to $t + \delta t$ are independent of the accelerations at $t + \delta t$. At any time step the initial displacements, velocities and accelerations are known and are used to calculate the final displacements. From these the final velocities and accelerations are calculated ready for the next step. Computation is relatively simple but the model can become unstable, producing nonsensical answers, if δt is larger than a critical value. This can be estimated or can be found by trial and error and is generally short so that many steps are needed. In *implicit analysis* the parameters are linked by more complex indirect relationships. In impact analysis, methods are called implicit if the displacements occurring during a time step cannot be found unless the accelerations are also known. The resulting equations are non-linear and are usually solved by linear approximations, of which several types exist. Many implicit methods are inherently stable regardless of the length of the time step, so that fewer steps are needed although the amount of computing at each step is greater. Which type of approach is the more efficient depends on the problem being considered. Explicit methods are probably more efficient when stress waves are important and conditions are changing rapidly, whilst implicit methods are probably more efficient if behaviour is slow enough for stress waves to be neglected. Explicit methods are more suitable for very simple models.

The use of complex computer models of structural response to impact is more of an art than a science. Skilled practitioners become very good at deciding which aspects are important and when things have gone wrong, but they are not infallible. Developed computer models can be bought and there are specialist firms which deal with the subject, but even the best models need to be checked against behaviour in real impacts. Protagonists of specific models are likely to be over-optimistic about their performance.

9.3.2 Single-mass models

The simplest model reduces a structure to a single mass attached to a single deflecting element. The analysis discussed in Section 7.2 can then be used. A major difficulty is deciding on the appropriate mass, its location and the characteristics of the deflecting element.

The equivalent system is usually selected so that the displacement of the mass is the same as that of a reference point on the structure. In the case of a beam this could be the mid point. If the mass of the structure is concentrated at the reference point then it can be used directly, but if it is distributed throughout the structure an effective mass M_e needs to be found. Usually the reference point is chosen to have a large displacement, and M_e is less than the actual mass because most of the structure is not displaced as far as the reference point (Biggs, 1959).

The effective mass can be found by making the kinetic energy in the model the same as that in the actual structure. It is necessary to assume a deflected shape for the structure, which could be that of the first mode of vibration or that of the structure with the dynamic loads applied statically. In the case of a simple beam of length l and mass m per unit length the kinetic energy is

$$T = \frac{1}{2} \int_0^l mv^2 \, dl = \frac{1}{2} M_e v_r^2 \tag{9.7}$$

where v is the transverse velocity at any point along the beam and v_r is the transverse velocity of the reference point. If m is uniform along the beam this gives

$$M_e = M \int_0^l v^2 \, dl / v_r^2 \tag{9.8}$$

The effective spring stiffness K_e for a vibrating system can be found from the natural frequency of the first mode, which is obtained by standard analysis methods. K_e is then found from

$$\omega = \sqrt{(K_e / M_e)} \tag{9.9}$$

The effective force F_e acting on the structure is found by making the work done on the model equal to the work done on the structure. With a simple beam of length l this is

$$W = \int_0^l f x \, dl = F_e x_r \tag{9.10}$$

where f is the transverse force per unit length, x is the transverse displacement at any point and x_r is the transverse displacement of the reference point.

If behaviour changes from elastic to plastic during the impact a separate model is needed for the plastic region because the deflected shape of the structure changes. Some simple examples are shown in Table 9.2.

Table 9.2 Equivalent spring mass system parameters for simple beams.

Loading case	Condition	Equivalent applied load Actual applied load	Equivalent mass Actual mass	Maximum resistance	Spring constant
Simply supported Concentrated load and mass at mid span	Elastic	1.0	1.0	$4M_0/L$	$48EI/L$
	Plastic	1.0	1.0	$4M_0/L$	0
Simply supported Uniformly distributed load and mass	Elastic	0.64	0.5	$8M_0/L$	$384EI/5L^3$
	Plastic	0.5	0.33	$8M_0/L$	0
Built-in ends Concentrated load and mass at mid span	Elastic	1.0	1.0	$8M_0/L$	$192EI/L^3$
	Plastic	1.0	1.0	$8M_0/L$	0
Built-in ends Uniformly distributed load and mass	Elastic	0.53	0.41	$12M_0/L$	$384EI/L^3$
	Plastic	0.5	0.33	$16M_0/L$	0

Based on Table 7.1 of Biggs (1959).

9.3.3 Lumped mass models

In a lumped mass model the structure is idealized as a number of concentrated masses connected by massless deflecting elements. The equations of motion of each mass can be derived, leading to a number of simultaneous differential equations which can be solved by standard techniques. In many cases the mass supported by the structure is concentrated in specific locations and a lumped mass model is easy to construct. In other cases the mass is distributed and needs to be approximated by a number of lumped masses. This calls for judgement and experience, and models may not be readily transferable from one establishment to another. The massless deflecting elements may be derived from test results on the appropriate components of the structure or their characteristics may be derived purely analytically. Appropriate dynamic magnification factors can be applied to static results to cover the effects of inertia and strain rate. Again different analysts may use different rules of thumb and models may not be readily transferable.

The deflecting elements in lumped mass systems can be arranged in different ways depending on the nature of the structure. Figure 9.20 shows a beam with two concentrated masses. In this case the masses are far coupled because a deflection of either mass causes a reaction at both supports. This may make no difference to the use of a model but needs to be borne in mind when constructing it.

There are limits to the extent to which it is worth subdividing the structure. With a small number of masses the deflecting elements are reasonably distinct from each other, but as the number of masses is increased it becomes harder to specify deflecting elements which are free from interaction with adjacent elements, and the model becomes less accurate than a simpler one. This depends on the nature of the structure. With motor car frontal crashes the upper limit appears to be about ten masses (Fowler and Newman, 1980), but models for crashes of large aircraft go up to eighty masses (Wittlin, 1983).

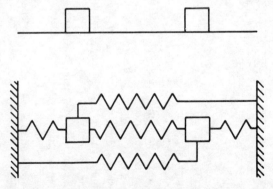

Figure 9.20 Lumped mass model of a beam carrying two concentrated masses.

Simple lumped mass models can be derived from first principles, but more complex models can be difficult to make and use and it is better to use commercially available computer programs or to employ a specialist computing firm.

9.3.4 Complex models

(a) Finite element models

In finite element models the structure is represented by a large number of interconnected elements. These vary in size, shape and nature and approximate to the characteristics of different parts of a structure. There appears to be no theoretical upper limit to the number of elements used, and complex models for structural impact may contain several hundred of them. A large range of elements appears in the literature and many are included in standard computer programs for structural analysis. Most of these are for small deflection static analysis, and impact modelling introduces additional complications because of dynamic effects, large deflections and the strain rate sensitivity of the material.

The amount of detail varies with the fineness of the finite element mesh. This can be varied over the structure so that areas of particular interest can be studied in more detail. Either complete structures or structural components can be modelled, and finite element results for components can be included in lumped mass models.

Finite element modelling of structural impact at other than a very simple level is not to be undertaken lightly. A comprehensive finite element package should include modular construction, so that some aspects can be altered without affecting the others, an extensive database, clear documentation and automatic program management. Provision of these and maintaining programs in good working order requires a large and continuous effort which is likely to be found only in specialist teams.

(b) Finite difference models

In finite difference models the equations for the behaviour of the structure are derived by treating the structure as a continuum. These equations are then solved numerically using approximations which divide them into discrete steps which are solved one at a time.

The distinction between finite element and finite difference methods often becomes blurred. Usually finite difference methods are used for fluid flow, which includes stress wave propagation and material behaviour under hypervelocity impact, whilst finite element methods are used for the static analysis of structures. In the analysis of the response of structures to impact it is often possible to use either method, the main difference being in the data management methods used in the computer programs. People will tend to use the method with which they are familiar.

Finite element methods are more convenient when components are irregular in shape, but such components can also be treated by finite difference methods at the expense of rather more complicated computing. Finite difference methods are more convenient when there are continuous smooth transitions in behaviour, but recent developments enable these to be analysed by finite element methods.

The non-expert can easily become confused by conflicting opinions, but in most cases it probably makes little difference which of the two methods is used.

10
Impact injury

10.1 INTRODUCTION

10.1.1 General

(a) Percentiles
People vary widely in many respects including response to impact. Generally these variations are expressed in terms of percentiles. If a man is fifth percentile in height it means that 5% of men are shorter than he is and 95% are taller. Percentile values vary with the population studied and, within a population, values for women differ from those for men. Values can also vary with age and the population as a whole can change with time (Damon, Stoudt and McFarland, 1966).

Percentile values can sometimes be added but only at the fiftieth percentile level. A man who was fiftieth percentile in the length of all body segments would also be fiftieth percentile tall, but a man who was fifth percentile in the length of all body segments would be appreciably less than fifth percentile tall. This is because a man who was fifth percentile tall would have several segments more than fifth percentile in length.

Ideally, in order to study the response of people to impact, appropriate percentile values should be chosen and, because fiftieth percentile values can be added, it is common practice to use them as a base. Often there is not enough information to specify percentile values and this leads to values being specified by qualitative statements such as 'not unreasonable' or 'reasonably conservative'. Where percentile values have been established the range can be very wide, with the ninety-fifth percentile resistance to impact injury being several times the fifth percentile value. This is much larger than the range for body dimensions of the British population where a man ninety-fifth percentile tall is only about 15% taller than a fifth percentile man and about 25% taller than a fifth percentile woman.

In simulating overall body response to impact a hypothetical person is often used who is fiftieth percentile in all body segment lengths and weights. Such a person does not exist and does not arise spontaneously from available data but has to be established by consultation and agreement. The internationally accepted reference population for this person is United States males as measured in the 1960s. When hypothetical fifth and ninety-fifth percentile men are considered they usually have the same proportions as the fiftieth percentile man adding up to the appropriate overall height and weight. Hypothetical women and children are also used but less frequently.

(b) Rules of thumb

The human body is very complex and its response to impact is not well understood. In some areas there is a large amount of information and in others there is very little. Even in areas which have been widely studied there is disagreement on the mechanisms of injury. Different mechanisms are possible, and these may change with the type of impact; the inherent variability in results makes it difficult to obtain unambiguous data; and it is not possible to investigate behaviour by tests on living people.

In the absence of adequate information a number of simplifying assumptions are widespread. Although these are usually logical they are not firmly based on either theory or empirical relationships and are best regarded as rules of thumb. They usually have a direct, intuitive appeal and they can be useful if they are not taken too seriously. Unfortunately their limitations are often not appreciated; they are treated as if they were both firmly based and precise, and complex analysis is often based on them.

Three rules of thumb are very common. The first is that some form of consistent cumulative damage rule applies, broadly analogous to Miner's rule in fatigue. Injury caused by a force–time pulse of one shape is then predicted from injury known to be caused by force–time pulses of other shapes by adding the times spent at different force or acceleration levels or by taking average levels regardless of pulse shape. The second is that the construction of the body is optimized in such a way that when one component fails a related component is close to failure. On this basis, for instance, the onset of other types of chest injury is indicated by the initiation of rib fracture. The third is that the body responds in the same manner to different types and durations of impact, so that a single dynamic model or a single empirical relationship can be applied to all impacts.

Such experimental evidence as is available indicates that these rules of thumb are all very rough approximations applying to very limited ranges of conditions.

(c) Sources of information

Information on the relationships between injury and impact has to be derived indirectly, and there are six main ways of doing this.

(1) Human volunteers are used to establish the likely lower bounds for injury. Most of this work is for military use and the subjects are fit young men wearing comprehensive restraint systems. The loads to cause a minor injury are likely to be higher than for the population as a whole.

(2) Clinical reports of injuries sustained during actual impacts establish the susceptible parts of the body and the nature and severity of the injuries. It is difficult to estimate the impact conditions, the injury mechanisms and the sequence of events.

(3) Cadavers are used in laboratory impact tests. When proper precautions are taken these appear to give useful results for bone fracture, but results for soft tissues are questionable. Cadavers tend to have been elderly people who were ill for some time before they died, and loads to cause a given fracture are likely to be lower than for the population as a whole.

(4) Anaesthetized animals are used in controlled impact tests and then killed and examined for signs of damage. There are differences in structure and size between people and animals which lead to difficulties in interpretation, but useful results can be obtained for soft tissue damage.

(5) Anthropomorphic dummies consisting of body segments connected by joints are used in controlled impact tests. Repeatability is poor by engineering standards but much better than for people. This can be improved by adhering closely to specified setting-up procedures but these may not represent realistic conditions. There are difficulties in relating measurements made on dummies to injuries sustained by people.

(6) Mathematical models of varying complexity are used. These can be divided into gross motion simulators, which usually model anthropomorphic dummies rather than people, and models simulating specific parts of the body such as the chest or spine. They are versatile and repeatable but can become very complicated and suffer from lack of reliable input data. There is also a tendency to assume that, because models give deterministic solutions, scatter has been eliminated from impact injury.

10.1.2 Basic relationships

(a) Tolerance limits

Levels of impact injury are often defined by tolerance limits. Ideally a tolerance limit specifies both the level of injury and the percentage of the population who would sustain this level of injury. In practice tolerance limits for impact injury are almost always based on inadequate data and there is usually not enough information to plot more than a very crude approximation to a tolerance limit.

It can be difficult to compare tolerance limits from different sources. Behaviour is affected by the use of a restraint system, its nature, the direction of impact and the posture of the person. It is also affected by where the impact is measured. Some tolerance limits are specified for impact directly on the body and some by the acceleration of the seat on which the person is sitting.

There are three fundamental tolerance limits: these define the maximum impacts that produce no injury, reversible injury and irreversible injury short of death. These can be subdivided to give a finer graduation or reduced to a single limit defining a specific criterion. For instance, a serious vehicle crash is potentially lethal and the criterion could be survival almost regardless of injury. With an escalator, on the other hand, it could be the minimum velocity change which would cause an elderly person to stumble.

(b) Assessment of injury

Injuries can be divided into three broad categories involving bones, soft tissues and the nervous system including the brain. Bones fail mainly by brittle fracture caused by the direct application of load. Fractures are usually apparent but damage to soft tissue is less clear cut. It can be caused by the direct application of load or by the relative displacement of organs within the body. There may be no outward sign of injury but it is usually visible on surgery or autopsy. Damage to the brain or nervous system may be deduced by impairment of function but there may be no signs of damage even on autopsy. When injuries of more than one type occur together the less tangible can be deduced from the more tangible, such as concussion from minor skull fractures, but there is not necessarily a good correlation among them.

Objective measurement of injury is seldom possible, and even such apparently unambiguous conditions as no injury or death can be interpreted in different ways. There are also possible long term effects such as unnoticed damage to the heart causing death several years later. Nevertheless, injury scales have been devised which rank injury on a descriptive or numerical scale. There appears to be a broad consensus over the application of these scales and it seems likely that, if data were available, fairly objective correlations would emerge (States, 1969).

10.2 WHOLE BODY BEHAVIOUR

10.2.1 Empirical relationships

The simplest analysis of whole body tolerance to impact just specifies a single value of acceleration as shown in Fig. 10.1 (Snyder, 1970, Fig. 19). Generally no indication is given of impact duration, the probability and level of injury or the body orientation. The accelerations are usually deduced from the velocity change, which is the only parameter likely to be known even approximately. Results for the highest accelerations are usually for accidental falls where survival has been unusual enough to be worthy of comment; almost all of the population are killed or badly injured. Results for the lowest accelerations are for fairly routine activities, such as parachute landings, where injury has been unusual enough to warrant comment; almost all of the population escape with little injury. Systematic study of accidental falls can provide appreciably more refined data but still largely on a case history basis.

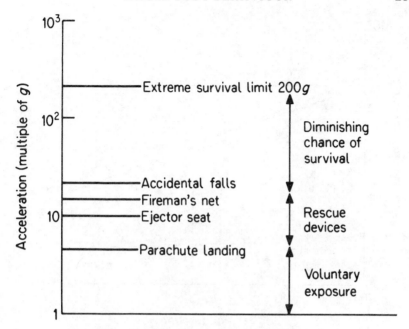

Figure 10.1 Whole body tolerance of impact.

At the next level of analysis the duration of the impact is given as well as the acceleration. A typical set of results is given in Fig. 10.2, which is for three levels of injury for a seated person restrained by a full safety harness (Versace, 1971, Fig. 2). The seat occupant was travelling forward and was rapidly decelerated so that acceleration is rearwards. The deceleration pulse is constant with an unspecified rise and decay time which are ignored. The pulse is applied to the seat and not the occupant. The results are mainly for tests on military volunteers and anaesthetized animals, with some estimates from accidental impacts.

There is an overall tendency for the level of acceleration to rise as the pulse duration falls, and there appear to be three regions: a short duration region for pulses less than about 0.05 s; a long duration region for pulses longer than about 0.1 s; and an intermediate region. The graph is plotted on logarithmic scales, and it is possible to draw straight lines approximately through the middle of the moderate injury band. A plausible straight line has the simple equation

$$TA^{2.5} = 1000 \tag{10.1}$$

where T is the pulse duration in seconds and A is the acceleration in multiples of g. Because of its simplicity this equation has been widely used, often out of context.

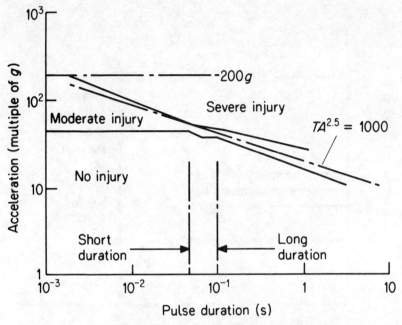

Figure 10.2 Whole body tolerance of rearward acceleration.

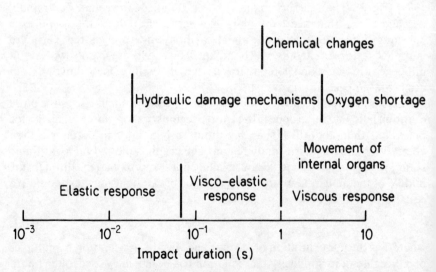

Figure 10.3 Suggested changes in body response with impact duration.

It seems unlikely that a single mechanism produces injury over the entire range, and some of the variations which have been suggested are shown in Fig. 10.3 (Gögler *et al.*, 1977, Figs 2, 3).

10.2.2 Dynamic models

(a) Spring–mass models

The body and its component parts have a number of resonances which have been studied at low amplitudes with steady-state sinusoidal inputs. Results for whole body behaviour are inconsistent, largely it appears because different people have different resonant frequencies. There is general agreement on only

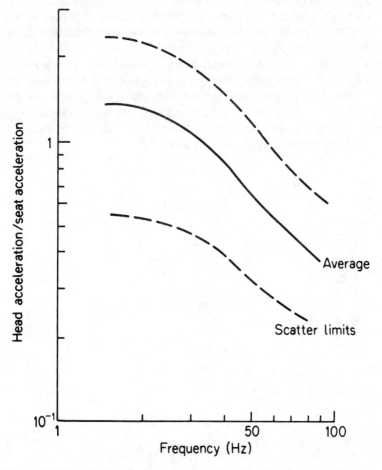

Figure 10.4 Variation in head acceleration in vertical whole body vibration.

one major body resonance, which is the fundamental mode for seated subjects vibrated vertically. This occurs at about 4 to 8 Hz.

Figure 10.4 shows how the ratio of head acceleration to seat acceleration varies for this type of vibration. Seat acceleration is measured at the interface between the occupant and the seat. With low amplitude vibrations, posture has a marked effect, and Fig. 10.5 shows how the ratio of head acceleration to seat acceleration varies with posture. Wide variations of the type shown in Figs 10.4 and 10.5 appear to be typical of the response of the body to vibration (Guignard, 1971).

Sometimes the body is represented by a simple spring–mass system with the characteristics shown in Fig. 10.6 (Glaister, 1978, Fig. 3). This is a version of Fig. 6.12 for response to a symmetrical triangular force–time pulse plotted in an alternative way. With vertical impact of a seated occupant, a dynamic overshoot of the type shown in Fig. 10.6 has been detected as shown in Fig. 10.7. This type of impact has been studied extensively because it occurs in aircraft ejector seats (Glaister, 1978, Fig. 5).

In other cases an overshoot has not been detected and the approximation shown in Fig. 10.8 is sometimes used with the curve of Fig. 10.6 reduced to two straight lines. For short duration pulses a given amount of injury is assumed to be caused by a given velocity change regardless of pulse duration. For long duration pulses the same amount of injury is assumed to be caused by a constant acceleration regardless of pulse duration.

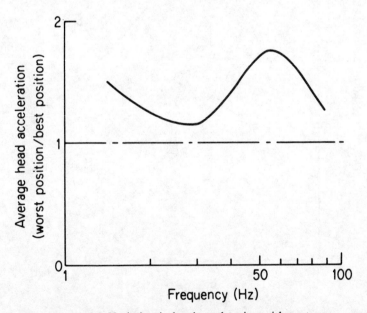

Figure 10.5 Variation in head acceleration with posture.

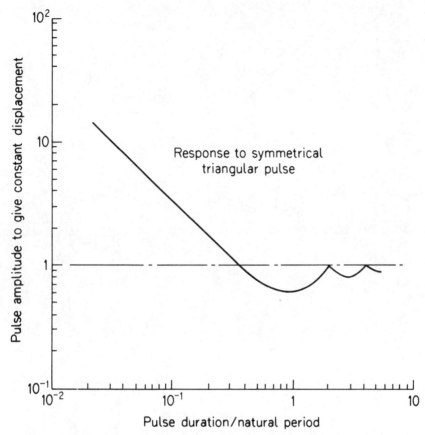

Figure 10.6 Response of body represented by a simple spring–mass system.

Fig. 10.9 shows the tolerance limit between injury and non-injury plotted in this way for the impact conditions shown in Fig. 10.2. For comparison the line $TA^{2.5} = 1000$ from Fig. 10.2 is superimposed. There is fair agreement between the two approximations close to the transition pulse length but increasing divergence away from there, illustrating that a single model or a single empirical relationship is unlikely to hold over a wide range of impact durations.

Tolerance limits represented in this way can be specified by three parameters – the velocity change, the peak acceleration and the pulse duration at which the two lines intersect. Table 10.1 gives these parameters for a number of cases.

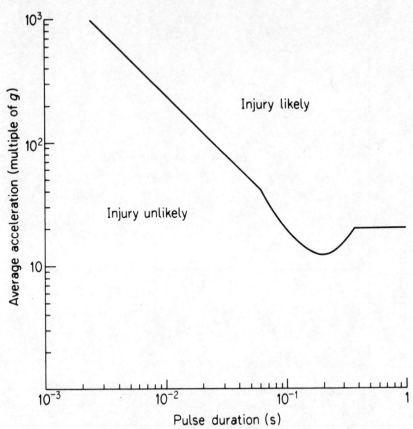

Figure 10.7 Response of occupant of an ejector seat which is subjected to a rectangular force–time pulse.

(b) Articulated rigid models

Unlike the previous analyses, which are used to codify existing information, articulated rigid models are used to assess behaviour in specified impacts. There are two basic types of model: anthropomorphic dummies, which are used in impact tests; and mathematical models, which almost invariably simulate dummies rather than people. There are three reasons for this. Firstly, there is a common problem of selecting a hypothetical person; secondly, it is easier to obtain measurements with dummies; and thirdly, dummies can be used in a wider range of test conditions.

The most commonly used size of hypothetical person is a fiftieth percentile male but larger and smaller adult males are sometimes used as well as women and children, where the proportions are different. Mathematical models are

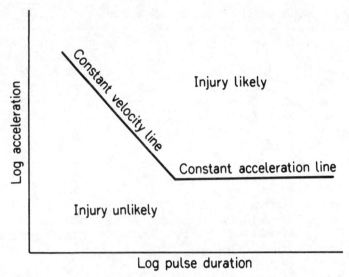

Figure 10.8 Two-line approximation of response to an acceleration–time pulse.

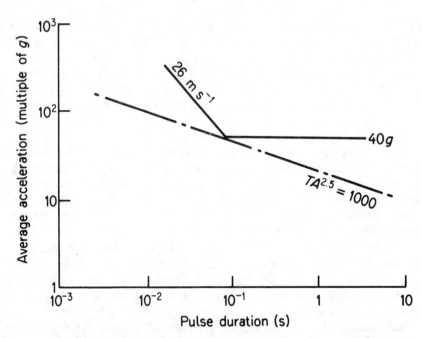

Figure 10.9 Response of fully restrained seat occupant to a rectangular pulse of rearward acceleration.

Table 10.1 Tolerance limits for whole body acceleration

Direction of applied force	Posture and restraint system	Velocity change (ms^{-1})	Constant acceleration (g)	Transition pulse duration (s)
Front to rear	Seated, full harness	22	40	0.06
Front to rear	Seated, lap and shoulder belt	8.5	20	0.04
Front to rear	Seated, lap belt	8.5	15	0.06
Front to rear	Seated, holding steering wheel	4.25	4	0.1
Front to rear	Standing	0.7	0.1	0.7
Transverse	Seated, lap and shoulder belt	4.25	8.0	0.06
Foot to head	Seated, lap and shoulder belt	22	20	0.18
Foot to head	Standing	3	2.5	0.12

Based on Glaister 1978

usually deterministic, giving unique repeatable results. They are sensitive to small changes in parameter values and the actual values chosen are likely to be arbitrary and may vary from user to user. Dummies do not give repeatable results because random changes in parameter values occur, even in closely controlled tests. In one experiment considered here, five different dummies were tested in one laboratory using skilled staff and identical equipment and test procedures. The variability is thus likely to be as low as could be obtained with those dummies (Chandler and Christian, 1969).

For all of the dummies the overall pattern of behaviour was similar, with the best fit value of the maximum seat belt load increasing linearly with the maximum seat deceleration. With individual dummies 95% of the results were within $\pm 10\%$ to $\pm 20\%$ of the best fit value. Over the five dummies the variation was much larger, with the seat belt loads found with one dummy being up to 2.5 times those found with another.

Variability did not seem to be correlated with dummy complexity or sophistication and it seems that, although greater complexity may give greater realism, it may also give greater scatter. Dummy characteristics can also change with wear, unobserved damage and the replacement of components known to be damaged.

Dummies usually have metal skeletons and rubbery soft tissue. They are stiffer and less highly damped than people and their dynamic response is usually quicker and higher. Some high frequency ringing usually occurs on impact and this can be filtered out, but other peculiarities of dummy response are harder to assess. Figure 10.13 shows an attempt to allow for differences in upper leg response between dummies and people (Viano, 1977).

The equations of motion for articulated mathematical models are usually derived by Lagrangian dynamics. Models can be two or three dimensional and of varying degrees of complexity. Results are usually presented pictorially, sometimes in three dimensions, with shading and a number of colours. Complex models take several years to develop and can be cumbersome and expensive to use. Even so they can be quicker and cheaper than impact tests using dummies (Wismans *et al.*, 1982). A major problem is lack of adequate data on the performance of the body under impact loading.

Two-dimensional models become inadequate when oblique impacts occur or asymmetrical restraint systems are used. The more complex two-dimensional models and most of the three-dimensional ones have two basic components – a body dynamics model and a contact model. The body dynamics model provides the overall motion of the body segments, and the contact model establishes when part of the body touches a surface. It also gives the force at the contact point. Models are usually validated against tests on anthropomorphic dummies. In view of the large number of fairly poorly controlled variables in the tests it can be difficult to check the performance of the model. Usually displacements look similar but forces and accelerations do not agree very well.

10.3 SPECIFIC BODY COMPONENTS

10.3.1 The legs

(a) Properties of leg bone

Impact injuries to the arms do not seem to have been studied in detail but there is a moderate amount of published information on the legs. This is almost entirely confined to the leg bones.

Bone can be regarded as a structural material and tested by standard techniques for the strength of materials. Static stress–strain curves for tensile and compressive specimens of fresh femur bone are shown in Fig. 10.10 (Yamada, 1970, Table 3). The specimens come from a limited age group of one population and the scatter in the results is small. Bone material from people and animals, alive or dead, is broadly similar although there are differences among species. Wet bone from newly dead bodies is very similar to living bone but dry or embalmed bone is stiffer and more brittle. There does not seem to be a standard definition of either dry or embalmed bone, and so details of the variation are not clear.

Low velocity impact tests using a standard pendulum testing machine have also been performed on specimens of fresh femur bone. The scatter in impact

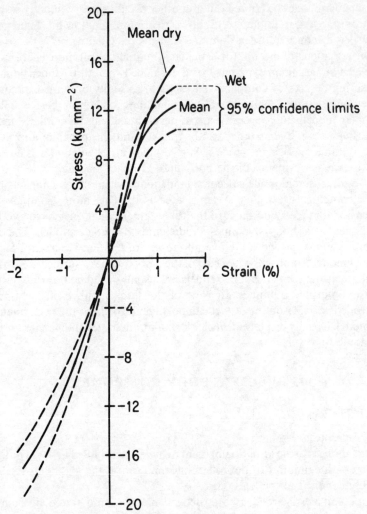

Figure 10.10 Tensile and compressive stress–strain curves for small specimens of femur bone from a single population and age group.

test results is much larger, as can be seen in Fig. 10.11 (Currey, 1979, Fig. 1). Both the ultimate tensile strength and the energy absorbed during impact fracture decrease with the age of the person. The strength of bone also decreases in some types of illness. Cadavers used in impact tests are often those of old people who were seriously ill for some time before they died, and their bones may be only about one-third as strong as bones from healthy young adults.

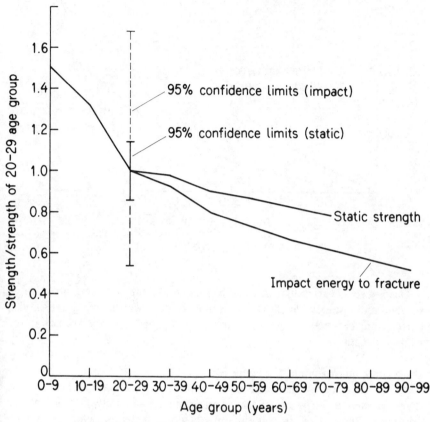

Figure 10.11 Variation in bone strength with age.

Another aspect of behaviour which has been studied is the effect of high velocity projectiles simulating bullets. The energy absorbed by the bone varies with the velocity and diameter of the projectile but does not seem to be affected by changes in the projectile mass. Some results are shown in Fig. 10.12. At the lower velocities the projectile makes a neat hole, but at high velocities the bone completely disintegrates in the region of the impact (Harger and Huelke, 1970).

Bone material is not homogeneous or isotropic and is sensitive to strain rate. In addition, although the material may be similar in different people, complete bones vary greatly in size and shape. Variations in posture, joint details and the size and shape of connected bones also occur, and similar overall loads can produce significantly different stresses in the bones of different people. The natural frequencies of specific bones have been found to vary by a factor of two, and this can be further increased by variations in the relative amounts of bone and soft tissue.

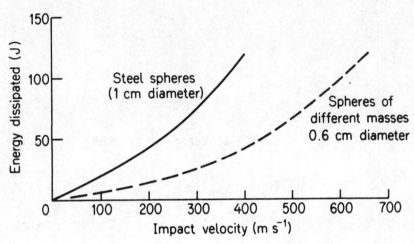

Figure 10.12 Penetration of femurs by projectiles.

Despite such variability, fracture occurs consistently at the highest stress reached and appreciable effort has been directed towards trying to establish correlations between impact loads and bone fracture. This is discussed in the next two sections.

(b) Axial compression of the upper leg

Axial compression of the upper leg is common in vehicle accidents, because there is often a fairly stiff piece of structure immediately in front of the occupants' knees. The bones mainly involved are the femur (which runs from the knee joint to the hip joint), the knee cap and the pelvis. Fracture can occur in any or all of these and the complete system is usually considered as a whole.

Fracture under axial impact has been studied directly in tests on cadavers and indirectly by reproducing, in controlled impact tests, the damage caused to the vehicle structure by the knees. The latter method can be used when measurable permanent deformation occurs to the structure and the load increases continuously with increase in permanent deformation (Lister and Wall, 1970).

The mechanics of the fracture process are complicated. The force–time pulse applied to the knee partly accelerates the upper leg as a whole and partly causes compressive loading in the bones. Behaviour depends on the interaction between the knee and the loading surface and on the relative masses of flesh and bone. The femur bends as a dynamically loaded strut and fracture occurs when the limiting tensile stress is reached in bending. The peak load on the knee usually occurs before the peak bending stress in the femur and can cause fracture of the knee cap instead or in addition. An overall statistical correlation

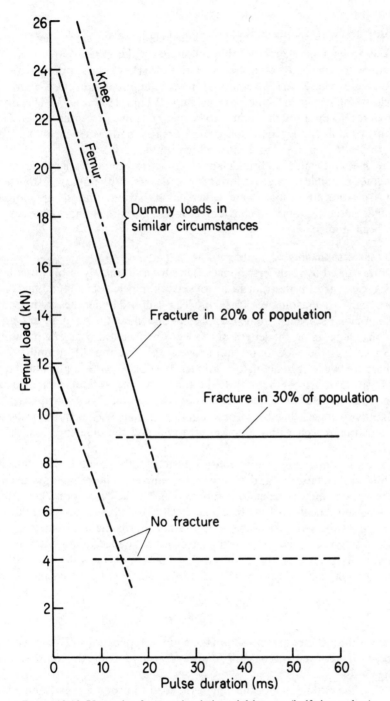

Figure 10.13 Upper leg fracture loads in axial impact (half-sine pulses).

can be found between the knee load and the likelihood of fracture, but there is unlikely to be a simple relationship in any particular case.

Empirical results have been used to derive the straight line approximations in Fig. 10.13. These are for a smooth symmetrical force–time pulse, approximately triangular or half-sine wave in shape (Viano, 1977, Fig. 3). There are enough results for a rough estimate to be made of some percentile values. The lines are broadly similar to the straight line approximations shown in Fig. 10.8 but are plotted on linear and not on logarithmic scales.

The knee and femur loads in an anthropometric dummy were measured in impact tests similar to these performed on cadavers. With impacts lasting less than 20 ms the dummy loads were higher and of shorter duration but followed a similar pattern. Dummy behaviour at longer durations does not seem to have been studied.

(c) Transverse loading

Fracture caused by transverse impact seems to occur mainly in the lower leg. Typical cases are a pedestrian struck by a car bumper, a skier stumbling and an occupant of a crashing vehicle whose leg is trapped under the seat in front. Bending can arise from a moment applied to one end of the leg or a transverse load anywhere along the length of the leg. There can also be complicating factors such as the effect of friction between the foot and the ground.

There are two long bones in the lower leg, the tibia and the fibula, which run roughly in parallel from the ankle to the knee. The tibia is about seven times as strong in bending as the fibula, which is usually ignored. The tibia appears to be slightly stronger when bent from front to rear than when bent from side to side, but this is well within the scatter found and the difference is usually ignored.

There are a small number of published results for the strength of the tibia in bending, both statically and in low speed impact. These are reported in different ways and, in order to compare them in the discussion below, the following simplifications have been made. All of the tibias are assumed to be fiftieth percentile with a length between joints of approximately 0.4 metres, and they are assumed to be simply supported beams loaded by a concentrated transverse load F. It is also assumed that fracture occurs when a constant maximum bending moment M is reached. Simple bending theory then gives

$$F = \frac{0.4M}{x(0.4 - x)}$$

where x is the distance in metres from the point of application of F to the end of the beam. It is convenient to take F as always positive, giving the curves shown in Fig. 10.14.

Within the middle half of the length the load to produce a given bending moment remains approximately constant, but in each outer quarter the load

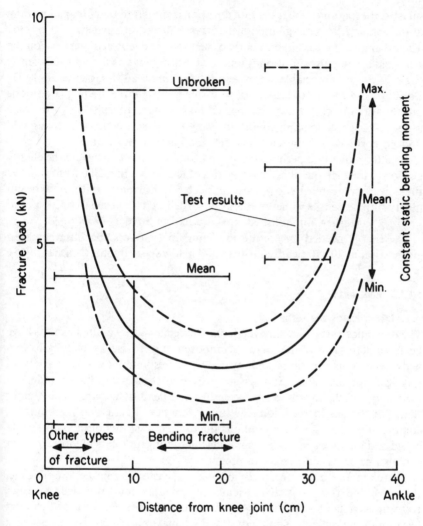

Figure 10.14 Fracture of tibia.

rises rapidly to infinity. Test results appear to support this, with fracture in bending if a transverse load is applied to the middle half of the tibia but some other type of fracture if it is applied closer to the knee. Three values of M are shown corresponding to minimum, mean and maximum values reported for skiing accidents (Hull and Allen, 1981).

Two other groups of impact test results are also shown in Fig. 10.14. One group is for impacts at unspecified locations between the mid length of the tibia and the knee. Fracture occurred in 20 to 30 ms, which can be regarded as quasi-static loading if the transition duration is similar to that for axial loading at the femur (J.W. Young, unpublished work quoted in Snyder, 1970, p. 736). The other group is for impacts fairly close to the mid length of the tibia on the ankle side. The pulse duration is unknown but it looks as if it was sufficiently short to cause appreciable increase in the fracture load (Kramer *et al.*, 1973). Both of these groups of tests were performed with simple loading conditions but the way in which they are reported makes them difficult to use. In some other tests, where loading conditions were complex, the investigators could find no correlation between transverse load and bone fracture.

An alternative, and perhaps more useful, approach is to measure the energy needed to fracture the tibia. In one series of tests more than 300 fresh cadaver tibias were fractured by low speed impact at mid length. Fifth, fiftieth and ninety-fifth percentile values of energy absorbed were reported to be 34, 66 and 114 J (B.S. Mather, unpublished work quoted in Snyder, 1970).

The energy needed to fracture the femur in transverse bending has been given as about 30 J statically and 40 to 50 J in low speed impact (Mather, 1968).

10.3.2 The torso

(a) Dynamics of the torso

The dynamics of the torso are very complicated and only limited aspects have been studied. Most of the torso is composed of soft tissues which perform a wide variety of functions and exhibit a correspondingly wide variety of physical characteristics. Soft tissues are visco-elastic, non-linear and non-isotropic, and the differences between live and dead tissues are much greater than for bone. In addition the lungs contain fluctuating amounts of compressible gas, the heart contains fluctuating amounts of incompressible liquid and the digestive system contains fluctuating amounts of food. Women can also be at varying stages of pregnancy.

There are four bone structures in torso – the shoulder girdle, the rib cage, the spine and the pelvis. It is usually regarded as good practice to design restraint systems so that loads are applied to bones. Because of this, and because of the difficulties in analysing the behaviour of soft tissue, tests and analyses have tended to concentrate on the bones. The shoulder girdle and the pelvis can be treated in isolation in the same way as the leg bones, but the rib cage and the spine interact strongly with the adjacent soft tissues. Analysis of chest behaviour has been largely empirical but a number of sophisticated mathematical models have been made of the spine.

The internal organs are not attached directly to the bones but are indirectly linked by other soft tissues. They are all contained within a single flexible enclosed space; thus load applied in one location can cause injury at a remote location, and there appear to be a number of different opinions about the major mechanisms of injury.

(b) Properties of soft tissue

There appear to be wide variations in published results for the mechanical properties of soft tissue, but the overall patterns are clear. In static tensile tests, soft tissues have three-stage stress–strain curves of the type shown in Fig. 10.15. At low stress there is a region of low stiffness where strain varies approximately linearly with stress; at intermediate stress the stiffness increases so that strain increases more slowly than stress; and at high stress there is a region of high stiffness where strain again varies approximately linearly with stress. The relative importance of the three stages varies (Yamada, 1970, Fig. 71).

Most muscle tissues fracture at static tensile stresses of about 0.1 to 0.3 N mm^{-2} and strains of 50% to 100%. Tendons are about 200 times as strong as this but can fracture at strains as low as 10%. Liver is only about one-tenth as strong as muscle, giving an overall range of 10^3 for the static tensile strengths of soft tissue. Some flexible membranes fracture at strains of 200% or more, giving an overall range of about 20 for strain at fracture.

Behaviour is visco-elastic and is broadly similar to that of rubber. Static tensile behaviour varies with previous loading; there is appreciable hysteresis, and creep and stress relaxation occur. Damping characteristics have been measured for small amplitude steady-state vibrations and various types of visco-elastic model have been proposed. In these models damping varies with frequency of vibration, but as with rubbers the actual damping is substantially independent of frequency over a wide range.

Behaviour is markedly anisotropic and analysis tends to be confined to one dimension. There is also some doubt about the relevance of tests on small

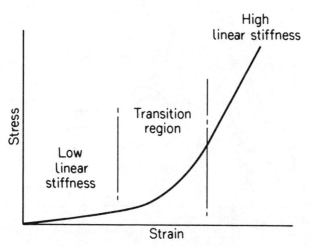

Figure 10.15 Typical static tensile curve for soft tissue.

isolated specimens to the behaviour of soft tissue in the living body. Cadavers do not have muscle tone and the soft tissues become stiff, so there are further doubts about their dynamic response.

(c) The spine

The spine contains twenty-four vertebrae which are separated by flexible discs and enclose the spinal cord. Although the construction of the spine is similar along its length, different sections of it (or even individual vertebrae or discs) can be studied separately. In particular the behaviour of the head and neck are closely linked and are discussed together in Section 10.3.3.

Various types of injury can occur under impact loading, but fracture of one or more vertebrae is generally taken as indicating the onset of serious injury. The way in which such fractures occur is fairly well understood and mathematical models of spinal behaviour under impact appear to be realistic, although some of them are rather complicated. Simple models of the type discussed in Section 10.2.2 do not take the details of spinal behaviour into account and are not adequate for detailed analysis.

Much of the study of spinal injury under impact loading has been concentrated on the axial compression which occurs in aircraft ejector seats. The spine is curved and the mass of the torso is offset from the axis of the spine, so even this type of loading produces appreciable bending as can be seen in Fig. 10.16. If a restraint system is used which limits the amount of bending, the tolerance limits of the spine can be raised appreciably, perhaps by as much as 70% (King, 1972).

Injury can also occur when a restrained seat occupant is decelerated at right angles to the spine. This has not been studied nearly as much as ejector seat

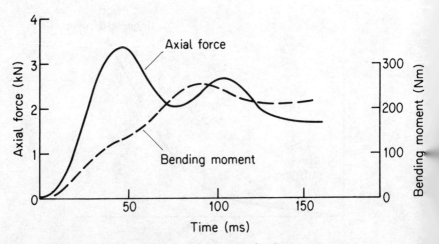

Figure 10.16 Loads on spine in ejector seat.

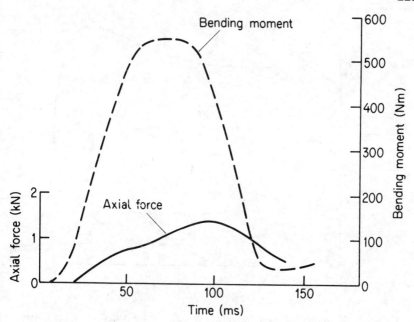

Figure 10.17 Loads on spine in car seat with lap and shoulder belts.

loading, and it is uncertain whether this type of injury will become a problem with the increased use of restraint systems. Some injury occurs when the torso bends violently round a lap belt, but fractures of vertebrae can occur even when the shoulders are restrained as well. Figure 10.17 shows compressive loads in the spine of a cadaver restrained by lap and shoulder belts. The loads in this instance are almost half the compressive loads for an ejector seat shown in Fig. 10.16, even though the shoulder belts are arranged so that they should not compress the spine (Begeman *et al.*, 1973, Fig. 14).

(d) The chest and abdomen

The dynamics of the chest and abdomen are poorly understood. There appear to be three different mechanisms which take different times to occur. In impacts lasting less than about 0.1 s the response seems to be largely elastic or visco-elastic, and spring–mass models are probably appropriate. In impacts lasting between about 0.1 and 2 s viscous flow of body fluids seems to be important, and in impacts lasting longer than 2 s chemical changes seem to start. Chemical effects can continue after the impact is over and it can be difficult or impossible to measure them.

The deflection of the chest has been studied because the rib cage spreads the load and upper torso restraint systems can act, at least in part, directly on the chest. Chest deflections can be large, being up to about 70% of the original

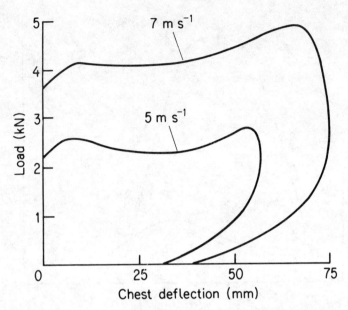

Figure 10.18 Smoothed average values of cadaver chest deflection on impact.

depth. Typical load–deflection curves for cadavers are shown in Fig. 10.18 (Neathery, 1974, Fig. 1). Deflection occurs at roughly constant load and there is little initial recovery, though presumably this takes place eventually in living people. The force generated may have three components – the spring of the rib cage, the compression of air in the lungs and the viscous or visco-elastic behaviour of the soft tissues. There may also be indirect contributions from attached masses such as the shoulders and arms.

Fracture of the ribs has been taken as indicating approximately the impact tolerance limit of the chest. This is a very imperfect measure because a young person with strong flexible ribs, receiving a load distributed uniformly over the rib cage, can suffer fatal chest injury without rib fracture. On the other hand, elderly cadavers with load concentrated on a few ribs can suffer rib fracture with no indication of other damage to the chest.

With a flat hard disc of 150 mm diameter impacting the front of the chest, rib fractures occurred in elderly cadavers at about 4 kN. At the other extreme, well-distributed loads of up to 13 kN have been reported without rib fracture; it has been suggested that a distributed load of about 6 kN might be a reasonable upper value to aim at in low speed impact of the chest. With more concentrated loads an impactor of 650 mm² caused fracture of individual ribs at 0.8 kN (Snyder, 1970, pp. 733–4).

A major effect of the rib cage seems to be that it distributes load before it gets to the internal organs. Lower torso restraint systems should act on the pelvis

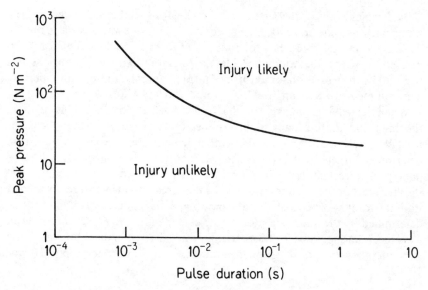

Figure 10.19 Response of chest to triangular blast pulse with zero rise time.

and not directly on the abdomen. It has been estimated that a standard seat belt acting across the pelvis could apply a distributed load at about 7 kN with a good chance of survival, but that the same belt acting directly on the abdomen would cause fatal abdominal injuries at a distributed load of 3.5 kN. With very well-distributed loads it has been suggested that the abdomen could withstand 5.5 kN from front to rear and 4 kN from side to side.

Damage to the lungs caused by blast waves from explosions has been studied, and Fig. 10.19 shows some results for probability of damage from a sharp-fronted triangular blast pulse. If the pulse lasts for less than about 0.05 s the pressure to cause damage starts to rise rapidly because the chest has not enough time to respond. Similar curves can be obtained for rapid decompression where a pressurized chamber fails (Von Gierke, 1964, Fig. 12).

(e) The shoulder girdle and pelvis
The shoulder girdle and pelvis can be regarded as basically bony structures similar to the legs but appreciably more complicated. The behaviour of the arms has not been studied as has that of the legs, and the dynamics of the shoulder girdle are complicated so that only a few approximate values appear to be available. Fractures of the collar bone were found in fresh cadavers at about 1 kN and fracture of the shoulder blade has been found at about 20 kN, leading to a suggested upper load limit of about 9 kN. It is difficult to apply loads specifically to the bones and the load paths are not clear. Loads are roughly as they would be applied by a safety harness. In the case of the

shoulder blades the harness passes under the arms so that it appears to be a lifting harness (Snyder, 1970, p. 734). For motor car side impacts a side load of 7 kN has been suggested as a tolerance limit for the shoulder. This has been derived from a combination of accident studies and tests on dummies, so it is the load as measured on an anthropomorphic dummy and not on a person. The comments in Section 10.3.1(b) apply (Harris, 1976).

When the pelvis is loaded from the front it is usually by an axial load along the thigh, and limiting loads are determined more by the femur than the pelvis itself. When is it loaded from underneath or from the rear the limiting loads are determined largely by the spine. The pelvis appears to be most susceptible to damage by direct contact when it is loaded from the side, causing fracture of the iliac crest or the hip joint socket. The iliac crest can also be directly loaded from the front by a seat belt, but the limiting loads in this case are likely to be determined by the abdomen. For motor car side impacts a tolerance limit of 6 kN has been suggested for the pelvis. This was derived in the same way as the side tolerance limit of the shoulder and the same comments apply (Harris, 1976).

10.3.3 The head and neck

(a) Dynamics of the head

Study of head dynamics has been almost entirely concentrated on brain damage, which is a major cause of death and injury in accidental impacts. Such damage is often not directly detectable and has to be deduced from various types of malfunction. As a result, although there are well-established medical procedures for establishing the presence of brain damage, there is considerable disagreement over the damage mechanisms involved. The problem has been studied widely, there is a large and growing literature, but no clear-cut answers have yet emerged (Goldsmith, 1981).

The head can be subjected to a direct impact or loaded indirectly by the neck. In either case head and neck injury are often closely linked. Three underlying injury mechanisms of closed head injury appear to be accepted widely, although there are disagreements over their relative importance. The first is linear acceleration of the skull causing pressure gradients in the brain, displacement of the brain relative to the skull, leading to both contact and cavitation damage, and displacement of the brain relative to the spinal cord. The second is rotational acceleration of the skull, leading to similar effects but with rotary rather than linear motion. The third is bending or stretching of the spinal cord where it joins the brain, where the skull and spinal cord meet or within the neck. In addition the skull can be penetrated, leading to open head injury with direct contact of the brain.

The skull can be regarded as a deformable solid of complex shape and the brain can be regarded as a viscous fluid. Mathematical analysis is difficult because the two dissimilar materials need to be treated simultaneously. Some models make sweeping simplifications such as regarding the head as a fluid enclosed in a tube or a sphere. Others aim for greater realism and involve complex computer programs. There is usually considerable difficulty in validating mathematical models and sometimes validation is not possible (King and Chou, 1976).

(b) Linear acceleration

Linear acceleration has been the most widely studied aspect of head impact injury because it presents behaviour in the simplest possible form. The most widely used set of empirical data is the *Wayne State tolerance curve* (from Wayne State University, Detroit), which is shown in Fig. 10.20. This plots a single tolerance limit against effective acceleration and time. As the time increases the effective acceleration to produce injury decreases. The curve has many shortcomings but it remains the best overall data set available (Versace, 1971, Fig. 1).

There are two main schools of thought. The first believes that the available data and understanding are so incomplete that not even approximate tolerance limits can be specified. The second believes that current knowledge is sufficient to form an *ad hoc* basis for such limits which can be used at present and improved progressively in the future.

All the data in the Wayne State curve are poorly defined. The effective acceleration is taken to be the mean value of the acceleration but the definition has been altered since the curve was first produced. It is not closely defined and depends on subjective assessment. The tolerance limit lies between injury which is dangerous to life and injury which is not. No percentile values are known and the limit is specified by what the original investigators judged to be reasonable in the light of their own extensive experience. The data came from

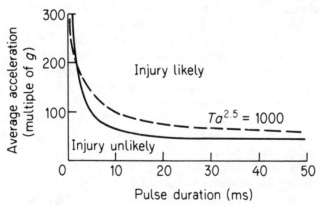

Figure 10.20 Wayne State tolerance curve for head impact.

three disparate sources. For short duration pulses the onset of fracture in cadaver skulls striking a flat rigid plate was used. For intermediate duration pulses, tests of a different type were performed on anaesthetized dogs. For the longer duration pulses, accelerations were estimated from events such as accidental falls and whole body decelerations of volunteers. The curve does not correspond to a consistent mathematical model and it seems likely that there are a number of different injury mechanisms acting, their relative importance changing with pulse duration.

In using the Wayne State curve it is assumed that the head is decelerated either by striking a hard object which is sufficiently blunt not to penetrate the skull, or via the neck as the body is decelerated. There are problems with deciding how the mean acceleration should be measured and derived. With direct contact of the head the skull vibrates, giving a dynamic amplification of acceleration which varies both with impact duration and the location of the measuring device. With loading via the neck, head acceleration depends on head mass and neck stiffness. Measurements are usually made on anthropomorphic dummies and the relationship of their accelerations to those of people are not clear.

The Wayne State curve can be fitted quite closely by empirical curve fitting or by assuming a simple dynamic model and adjusting its parameters to give the appropriate response. Because the data used in deriving the curve are so imprecise, accurate fitting does not seem to have been given high priority.

The most well-known use of the Wayne State curve was to derive the *head injury criterion* (HIC),

$$Ta^{2.5} = 1000 \tag{10.2}$$

where a is the mean acceleration of an applied acceleration time pulse in g, and T is the pulse duration in milliseconds (Johnson and Mamalis, 1978). Note that although Equation 10.2 is very similar to Equation 10.1 and the two are often confused, they are different. Equation 10.1 is the equation of a line defining a tolerance limit for whole body acceleration. Equation 10.2 is a weighted measure of the area under an acceleration time pulse used to define a tolerance limit for brain injury. The similarity between the two is an historical accident caused by some early misconceptions. If symmetrical triangular pulses are assumed applied to a simple brain model and the HIC is used in reverse to derive the Wayne State curve it does not give a good fit as can be seen in Fig. 10.20. A full discussion can be found in Versace (1971). Evidence accumulated since the Wayne State curve was first produced indicates that it is probably over-conservative at longer durations. In some applications the constant term on the right-hand side of the equation has been increased to 1500, increasing the tolerance level still further.

In applying Equation 10.2 to predicting head injury, measurements of acceleration are made on dummy heads. The area under the measured acceleration–time pulse is divided by the pulse duration to give the mean acceleration. This is then raised to the power of 2.5 and multiplied by the pulse duration to give the left-hand side of the equation. There are problems in defining the appropriate pulse length. If the start and finish time chosen are varied, the value of the head injury criterion changes, reaching a maximum for a specific start and finish time. Sometimes this maximum head injury criterion is used. There is an implicit assumption throughout that some undefined but consistent damage mechanism is at work, and the head injury criterion does not differentiate among different shapes of pulse.

The dynamic models used to simulate the Wayne State curve are lumped mass models with one or two degrees of freedom. The deflecting elements use linear springs and viscous damping. A decision has to be made on an appropriate injury mechanism, and usually maximum relative displacement between brain and skull is used (Fan, 1971). The models can be made to fit the Wayne State curve with varying degrees of accuracy, but otherwise they are difficult to validate and suffer from lack of adequate input data.

When direct impact to the head is likely, a helmet can be worn. The energy input to cause skull fracture is about 50 to 100 J with a blunt impactor. The force to cause skull penetration varies with impact speed and the size of the impactor. For an impactor with a cross-sectional area of about $100 \, \text{mm}^2$ it varies from about 1 kN in a static test on a weak skull to about 6 kN in a strong skull at an impact speed of $8 \, \text{m s}^{-1}$ (Snyder, 1970, pp. 728–30). The purpose of a helmet is to distribute concentrated loads across the whole skull and to absorb a significant amount of the input energy. There are a number of standard test criteria for helmets; the standards vary in detail from country to country but have a strong interrelation. They involve low speed impacts with energies of 120 to 160 J and forces in the helmet of about 15 kN (Sarrailhe, 1984). The helmet is fitted to a rigid head form and dropped through about 2 m to strike a rigid anvil. Not all of the impact energy is absorbed and the helmet rebounds. The peak acceleration of the head form is specified as about $300g$ which, as can be seen from Fig. 10.20, is at the lower duration end of the Wayne State curve, implying a pulse duration of about 1.5 ms. The head injury criterion is sometimes used in helmet testing, but often only a few simple measurements and visual checks are used (Newman, 1975).

A helmet consists of two main parts – a strong outer shell and an inner energy absorbing liner. The rigid head forms in use are much stiffer than the skull and are not regarded as satisfactory, although opinions on their effect on test results differ. It has been said that the shapes are incorrect so that the helmet does not fit properly, that the head forms greatly reduce deflection of the helmet shell, and that hard head forms can produce artificially low levels of head acceleration because they interact with the liner in the wrong way.

(c) Rotational acceleration

When the head is loaded via the neck there is always rotational acceleration and this can be found with direct impact too. There does not seem to be an overall empirical tolerance relationship, even one as approximate as the Wayne State curve, but evidence shows that rotational acceleration can be a significant factor in brain damage (Aldman *et al.*, 1982). Usually rotational and linear acceleration occur together and it is difficult to separate their effects. A number of mathematical models have been produced for skull, brain and neck response to rotational acceleration, but there seems to have been little attempt to relate these to empirical results (King and Chou, 1976).

(d) Flexure of the neck

The neck flexes forward readily, coming to rest when the chin touches the chest. Little damage seems to be caused by this in impact loading. It flexes sideways and backwards less readily and there is nothing to stop excessive flexing, so serious injury can occur. The most common type of injury is when a seated person is rapidly accelerated forward and, unless the seat has a head support, the neck bends backwards. The resulting injury can involve brain rotation, movement of the brain relative to the brain stem where it connects with the spinal cord, damage to the spinal cord and fracture of the neck vertebrae. This is known as *whiplash injury* (Snyder, 1970, p. 731).

Symptoms may appear immediately or not till several months later. Low speed impacts can produce as much injury as high speed ones, presumably because the head on the neck can act as a resonant system and displacement can vary with the shape of the acceleration–time pulse as well as its duration and the energy input. Given the wide variation in the natural frequencies of people, this would effectively give a random variation in injury regardless of impact speed. It is difficult and dangerous to study even low levels of whiplash on volunteers, and the only overall answer appears to be adequate head support to rear and sides if there is a chance of excessive neck flexure occurring.

11

Testing

11.1 INTRODUCTION

11.1.1 General

(a) Impact testing

Impact testing differs from most other types of testing in that corrections cannot be made once the test has begun. This means that pretest preparation and checking needs to be done very thoroughly and that results need to be checked to ensure that they make sense. The additional preparation and checking introduces a level of complication above that of a static test, broadly equivalent to introducing an additional dimension; for instance, an effectively one-dimensional impact test is about as complicated as an equivalent two-dimensional static test. The need for checking results is not always appreciated, and reputable journals have published plausible papers with results which are grossly in error. Very simple checks are sufficient, such as ensuring that the overall force, energy or momentum balance is correct.

The need for constant checking in situations which may not be well defined means that continuity of staffing is very important. Experienced staff can produce repeatable results but much of their skill is intuitive rather than codified, so that nominally similar tests performed at different establishments can give appreciably different results.

Impact behaviour is often poorly understood and there are usually a number of imponderables involved in a test. Because of this, much testing is *ad hoc*, looking at a limited set of conditions with a specific end in view. Experience gained in such testing may be very specific and there seems to be a lack of communication amongst people engaged in different types of impact testing.

In an impact test everything has to be done at the same time and, if the test is at all complex, there are a large number of things to do. There are two basic ways of dealing with this: the first is to have enough people present to operate all the equipment simultaneously, and the second is to have preset equipment which is triggered automatically. The first is labour intensive, flexible and subject to error; the second is capital intensive, less flexible and less subject to error. Which is used depends on the number of tests, the repeatability required and the type of test.

It may not be known whether the test has been successful until after the results are analysed and checked. Obvious faults such as a camera not working can be seen right away, but more insidious errors such as a malfunctioning transducer channel cannot. It is usually advisable to have the most important items recorded twice, either by duplicating the instrumentation exactly or by recording the item in a slightly different way. This is more important than adding extra channels of less important information.

A significant problem can be preventing further damage to the specimen after the test is over. This can be caused by the need to arrest a specimen which is still moving or by subsequent handling. The problem is more likely to arise with relatively flimsy specimens such as crashed motor cars or shattered concrete panels.

(b) Discrepancies in test results

A common feature of all forms of testing is that it is often difficult to separate the results obtained from the assumptions made and the methods of analysis used. Consequently many test results are semi-empirical and the relationships derived may not hold if the test conditions are altered significantly. Relationships in general use, which have been obtained in other types of test, may not be directly applicable to impact and should be checked before they are used.

Usually it is easier to measure displacements rather than velocities, accelerations or forces, but it is difficult to discriminate among different theories of behaviour using displacements alone. There is seldom unambiguous proof that one hypothesis is better than another and, as there is often appreciable scatter in test results, a number of different theories can be used to explain a given set of test data. Comparisons among different semi-empirical formulae may be difficult or impossible, and controversies arise because this inherent ambiguity is not appreciated.

11.1.2 Basic features

(a) Propulsion and guidance

Two essentials for a successful impact test are a means of bringing a moving mass up to speed and a means of ensuring that the impact occurs at the right location and in the correct orientation. In general, accurate guidance is more important than accurate speed because it is easier to make allowances for speed variation in a test which is otherwise correct than to compensate, for instance, for an event occurring partly out of the field of view of a camera.

The propulsion used varies with the impact speed, and some typical types of propulsion are shown in Fig. 11.1. At low speeds modified static test apparatus may be used. The speed achieved in such apparatus tends to increase with a decrease in the size of the specimen. Gravity provides a simple and repeatable

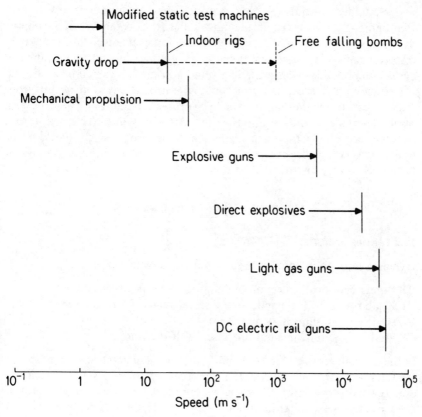

Figure 11.1 Types of propulsion.

means of propulsion at speeds up to about $25\,\mathrm{m\,s^{-1}}$, but the height of drop needed increases with the square of the impact speed. Mechanical or electro-mechanical propulsion is useful up to a maximum speed of about $50\,\mathrm{m\,s^{-1}}$. Above this compressed gas, chemical explosives or electromagnetic force can be used.

With small specimens used to test material properties, guidance is usually maintained throughout the test. Otherwise guidance is usually removed once the moving mass is up to speed. Sometimes an impact force such as that from an explosion is applied to a specimen directly without any intermediate stage, but usually the specimen is propelled towards a stationary target or a projectile is fired at a stationary specimen.

(b) Data retrieval

In most impact tests some residual permanent deformation is caused and this is often the easiest thing to measure. High speed photographs can extend understanding considerably by showing such deformations developing. Photographs can be taken during most impact tests regardless of the impact duration, the amount of deformation or the forces involved, but show only displacements. Electronic instrumentation is more limited in its range of application but displacement, velocity and acceleration can all be measured.

Extracting numerical data from photographs is tedious but records from electronic instrumentation can usually be processed simply and quickly. As a result electronic instrumentation is used extensively in lower speed impact tests but, as the impact speed increases, it becomes progressively more difficult to use. At very high impact speeds photography is used much more to obtain both quantitative and qualitative information.

11.2 TESTS OF MATERIALS

11.2.1 Aims of testing

Dynamic tests on materials are made for three main reasons.

(1) Purely empirical collection of data for a specific application
(2) Establishment of dynamic stress–strain relationships analogous to the static relationships in common use
(3) Study of the fundamental processes of deformation.

In the first category are the Izod and Charpy tests of notched specimens and the US Navy dynamic tear test. It is not necessary to understand or measure exactly what is occurring as long as repeatable results are obtained and test machines can be standardized and calibrated. In the second category it is necessary to eliminate or allow for the dynamic response of the test machine and perhaps the dynamic response of the test specimen so that characteristic behaviour can be established regardless of the details of the test used. In the third category a thorough examination and analysis of microscopic changes in the specimen is likely to be needed together with some additional measurements during the test.

The three types merge into each other and, for convenience, the second type is discussed. The other two types can be regarded as relaxing or tightening the procedures used.

11.2.2 Effects of stress waves

(a) Stress waves not important
At the lowest impact velocities, tests are basically extensions of conventional static tests. The test duration is long relative to the time taken by an elastic stress wave to traverse the test machine and stress waves can be neglected, although the dynamic response of the test machine needs to be allowed for.

Modified static test machines can often be used. Considerable ingenuity has been shown in the design of test machines but there seem to be no general rules unique to impact testing (Grundy *et al.*, 1985, especially pp. 34–5).

Usually the initial dynamic response is more important than residual vibrations, with care being taken to minimize dynamic overshoot on initial loading. It is also important to ensure that the location where load is measured is not moving out of phase with the location where displacement is measured. With electronic measurement and digital analysis it is possible to allow for the dynamic response of the test machine to obtain the true loads on the specimen.

The loading pulse can be achieved in various ways. Common ones are falling weights, rotating flywheels with fast acting clutches or cams and high pressure hydraulics with gas accumulators and electrically operated servo values. In general, falling weights give a specified energy input, cams give specified displacement–time pulses and hydraulic machines give specified force–time pulses.

Strain rates up to about $10^2\,\mathrm{s}^{-1}$ can be obtained. A specimen configuration which is attractive in theory but difficult to use in practice is the expanding ring or cylinder. This consists of a thin circular ring or cylinder of the test material which is expanded radially by a symmetrical force pulse (Zukas *et al.*, 1982, pp. 311–13). The pulse can be provided by an explosive charge, an exploding wire or electromagnetically (Forrestal *et al.*, 1980). If the pulse is of sufficiently short duration the ring can be regarded as loaded by an impulse which gives it an instantaneous outward velocity.

In the ideal case there are no circumferential stress waves travelling round the ring, which is loaded in uniform tension. After a very short time stress waves traversing the ring in the radial direction destroy this simple behaviour.

The behaviour of a small segment of the ring at time t after the initial impulse is shown in Fig. 11.2. The length of the segment is $l = 2\theta R$ for small values of θ,

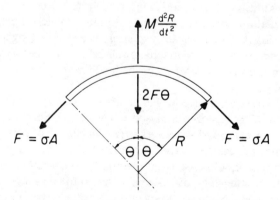

Figure 11.2 Force balance in segment of expanding ring.

and its mass is

$$M = 2A\rho\theta R$$

where A is the cross-sectional area and ρ the density. The tensile force in the ring is

$$F = \sigma A$$

where σ is the tensile stress. The component of force in the direction of motion is $2F\theta$.

Equating the forces in the element in the direction of motion gives

$$2\sigma A\theta = -2A\rho\theta R\frac{d^2 R}{dt^2}$$

$$\sigma = -\rho R\frac{d^2 R}{dt^2} \tag{11.1}$$

The problem in using this relationship is that R has to be differentiated twice to give σ, and it is difficult to get values of σ which are accurate enough to be useful. This can be improved by using a transducer, such as a Doppler laser, which gives velocity directly so that only one differentiation is needed.

If the ring is assumed to expand symmetrically and the strain is assumed to be uniform, then both strain and strain rate can be derived from the change in R.

The method has been used to a limited extent and strain rates up to $10^4\,\text{s}^{-1}$ have been reported. A ring gives plane stress in the specimen and a cylinder gives plane strain.

(b) Stress waves important in test machine

For any given strain in the specimen the test duration decreases as the strain rate increases until, at a sufficiently high strain rate, the test duration is similar to the propagation time of stress waves. Because the test machine is usually appreciably larger than the specimen, stress waves take longer to traverse it and become important in the machine well before they become important in the specimen. Elastic stress waves in the machine can then be utilized in the loading process.

The most common device in which stress waves are utilized is the *split Hopkinson bar*. This has been analysed extensively to check on the validity of the assumptions made and the ranges over which it can be used. It is generally accepted that, subject to suitable precautions, the device is accurate and reliable. The technique can be used for compression, tension, torsion, shear or bending but, for simplicity, discussion is limited to compression which is the simplest case. The loading in the specimen is plane stress (Zukas *et al.*, 1982, pp. 287–307).

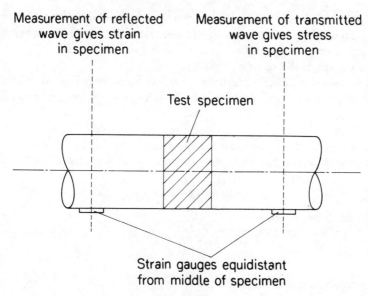

Measurement of reflected
wave gives strain
in specimen

Measurement of transmitted
wave gives stress
in specimen

Test specimen

Strain gauges equidistant
from middle of specimen

Figure 11.3 Split Hopkinson bar.

A split Hopkinson bar for compression testing is shown schematically in Fig. 11.3. The test specimen is placed between two long slender rods of a material which is linearly elastic and stronger than the specimen. Usually the test is arranged so that the specimen flows plastically. For simplicity it is assumed in the discussion below that the two rods and the specimen all have the same cross-section, but the analysis can be extended readily to cover different cross-sections.

A compression pulse is applied to the free end of one rod and an elastic compression wave travels along the rod until it reaches the specimen. The wave is then partly reflected at the interface as a compression wave increasing the compressive stress in the rod from σ_0 to

$$\sigma_1 = \sigma_0 + \sigma_r$$
$$= \rho C(v_0 + v_r)$$

where σ_0 and σ_r are the initial and reflected stresses, v_0 and v_r are the initial and reflected particle velocities and C is the wave front velocity.

In addition the wave is partly transmitted to the specimen. This produces a compression wave which travels to the far end of the specimen, where it is again partly reflected and partly transmitted. The transmitted part of the wave produces a compressive stress wave in the second rod of amplitude

$$\sigma_2 = \rho C v_2$$

The specimen is made much shorter than the loading rods so that stress waves traverse it much more quickly, and it can be regarded to a good approximation as being always at a uniform stress of $\sigma = \sigma_1 = \sigma_2$ so that $v_0 + v_1 = v_2$ if the rods are both of the same material.

If the specimen deforms plastically and is continually getting shorter, end 1 moves faster than end 2 and the drop in velocity along the specimen is

$$\delta v = v_1 - v_2$$

Because the end of the first rod is partly restrained by the specimen, the particle velocity in the rod is reduced when the stress wave is partly reflected, so that $v_1 = v_0 - v_r$. Hence

$$\delta v = (v_0 - v_r) - v_2$$

Substituting for v_2 gives

$$\delta v = (v_0 - v_r) - (v_0 + v_r)$$
$$= -2v_r$$

The strain rate in the specimen is

$$\dot{\varepsilon} = \frac{\delta v_r}{L}$$

$$= -\frac{2v_r}{L}$$

$$= -\frac{2}{L} C \varepsilon_r$$

where L is the specimen length and $\varepsilon_r = v_r / C$ by simple stress wave theory. Integrating both sides gives

$$\varepsilon = -\frac{2C}{L} \int_0^t \varepsilon_r dt \qquad (11.2)$$

Thus the strain in the specimen at any time can be derived from the strain produced in the first rod by the reflected part of the compression wave.

The stress in the specimen is

$$\sigma = \sigma_2 = E \varepsilon_2 \qquad (11.3)$$

and can be derived from the strain caused in the second rod by the transmitted compression wave. If strain gauges are attached to both at equal distances from the specimen the respective wave fronts reach them at the same time and the strain and stress in the specimen can be monitored simultaneously. When the compression wave reaches the free end of the second rod it is reflected as an unloading tensile wave which unloads the specimen and ends the test.

Strain rates up to $10^4 \, s^{-1}$ have been reported.

(c) Stress waves important in specimen

If the loading time is reduced still further relative to the time taken by a stress wave in traversing the specimen, the behaviour of stress waves in the specimen can be utilized. Two well-known cases are the Taylor cylinder, in which a short cylinder is impacted at up to about $300 \, \mathrm{m \, s^{-1}}$ to estimate the dynamic yield stress, and tests to determine shock wave behaviour in solids, where a flat plate is impacted at about $3000 \, \mathrm{m \, s^{-1}}$ (Zukas *et al.*, 1982, pp. 308–10; Goldsmith, 1960, pp. 186–93; Johnson, 1972, pp. 229–49).

In the Taylor cylinder the specimen is loaded in plane stress and the strain rate is about $10^3 \, \mathrm{s^{-1}}$. The specimen is fired at a flat rigid anvil and, at time t after initial contact, the conditions are as shown in Fig. 11.4. A plastic wave front travels at velocity C_p, which is assumed to be so much slower than the elastic wave velocity that elastic deformations can be assumed to occur instantaneously. Behind the wave front the cylinder is plastically deformed and stationary, whilst ahead of the wave front the cylinder is in uniform elastic stress and still moving.

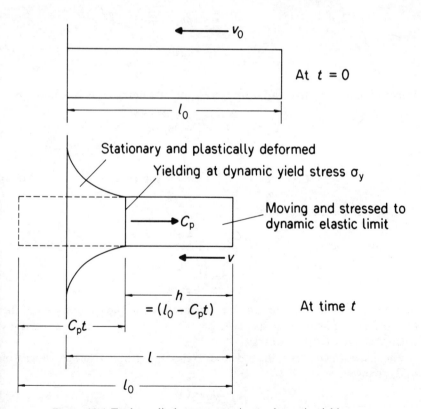

Figure 11.4 Taylor cylinder test to estimate dynamic yield stress.

The equation of motion of the moving part of the cylinder is

$$A_0 \sigma_y = - A_0 \rho h \frac{dv}{dt}$$

where A_0 is the original cross-sectional area, h the length of the moving part, σ_y the dynamic yield stress and ρ the density. This gives

$$\frac{dv}{dt} = - \frac{\sigma_y}{\rho h}$$

If the absolute velocity of the plastic wave front is C_p its velocity relative to the moving part of the cylinder is

$$C_p + v = \frac{dh}{dt}$$

$$\frac{dv}{dh} = \frac{dv}{dt} \frac{dt}{dh}$$

$$= - \frac{\sigma_y}{\rho h (C_p + v)}$$

Thus

$$(C_p + v) \, dv = - \frac{\sigma_y}{\rho} \frac{dh}{h}$$

Integrating from v_0 to v and from l_0 to h, and assuming C_p is constant, gives

$$\left(C_p v + \frac{v^2}{2} \right)_{v_0}^{v} = \left(\frac{\sigma_y}{\rho} \log_e h \right)_{l_0}^{h}$$

or

$$C_p(v - v_0) + \frac{1}{2}(v^2 - v_0^2) = \frac{\sigma_y}{\rho} \log_e (h/l_0) \tag{11.4}$$

At any time during the impact

$$C_p t = l_0 - h$$

giving

$$t = \frac{l_0 - h}{C_p} \tag{11.5}$$

Assuming uniform deceleration and constant mass of the moving part of the rod, and taking final values as $t = T$, $l = L$, $h = H$ and $v = 0$, gives conditions when the rod has come to rest.

The undeformed part of the rod has travelled $l_0 - L$ at an average velocity of $v_0/2$, giving

$$T = \frac{2(l_0 - L)}{v_0}$$

From Equation 11.5

$$T = \frac{l_0 - H}{C_p}$$

and eliminating T gives

$$C_p = \frac{v_0(l_0 - H)}{2(l_0 - L)} \tag{11.6}$$

Substituting in Equation 11.4 gives

$$-\frac{v_0^2}{2}\left(\frac{l_0 - H}{l_0 - L} + 1\right) = -\frac{\sigma_y}{\rho}\log_e\frac{H}{l_0}$$

or

$$\sigma_y = \rho\frac{v_0^2}{2}\left(\frac{l_0 - H}{l_0 - L} + 1\right)\log_e\frac{H}{l_0} \tag{11.7}$$

If it is further assumed that C_p is appreciably higher than v_0 then, from Equation 11.6, $(l_0 - H)/(l_0 - L)$ becomes appreciably larger than 1 and

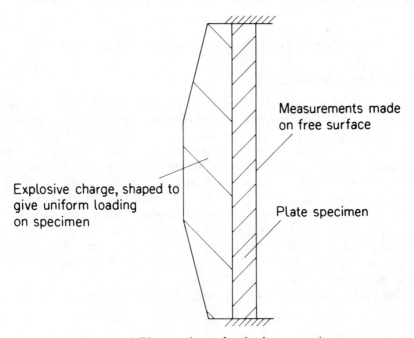

Measurements made
on free surface

Explosive charge, shaped to
give uniform loading
on specimen

Plate specimen

Figure 11.5 Plate specimens for shock wave testing.

Equation 11.7 tends to

$$\sigma_y = \rho \frac{v_0^2}{2}\left(\frac{l_0 - H}{l_0 - L}\right)\log_e\frac{H}{l_0} \tag{11.8}$$

From this, σ_y can be found by measuring l_0, L and H.

The test technique and the analysis can be refined to reduce the number of approximations, but the method suffers from the usual inaccuracies found in deriving details of behaviour from residual deformations. The results in Fig. 1.2(a) were obtained by this method.

Tests to determine shock wave behaviour use flat plate specimens impacted at velocities of about $3000\,\mathrm{m\,s^{-1}}$, giving strain rates of about $10^5\,\mathrm{s^{-1}}$. Loading is in plane strain and the impact is applied either directly by an explosive charge, as shown in Fig. 11.5 (Goldsmith, 1960, Fig. 76), or indirectly by a flat plate which is propelled by a direct contact explosive charge (Kinslow, 1970, pp. 310–39). Measurements are made of the movement of the other face of the plate specimen which is free. The shock wave velocity is derived from the time the wave takes to reach the free surface, and the particle velocity from the velocity of this free surface when the shock wave has reached it. Because the surface is free it is assumed to be moving at twice the particle velocity. The Hugoniot equations given in Section 2.4.1(c) are then used to derive the further parameters. The test technique and the analysis can again be refined. The Hugoniot relationships shown in Fig. 2.7 were obtained by this method.

11.3 TESTS OF COMPONENTS AND SYSTEMS

11.3.1 Aims of testing

Impact tests of components, or complete systems, are intended to simulate simplified and stylized impacts typical of those found in service. Tests of complete systems give overall behaviour and can sometimes highlight the importance of factors which had been neglected. They are likely to be appreciably more expensive than tests on components; thus fewer of them can be performed, making it difficult to decide on which conditions to select. It may also be difficult to study the behaviour of individual components and, with complex systems, results may be variable. Tests of subsystems or components are usually simpler and cheaper, and so a wider variety of conditions can be simulated and estimates can be made of the variability to be expected. The mounting of components can influence the test results and it may be difficult to simulate actual conditions.

Tests fall broadly into three categories:

(1) Acceptance or compliance tests, in which the specimen passes or fails a specified criterion

(2) Development tests, which are part of a test programme that can range from entirely empirical to largely theoretical with a few check tests
(3) Research tests, which may aim to establish the tests to be used in the other two categories or to study the fundamental process involved.

These are likely to be similar in execution and there are no rigid boundaries dividing them. Acceptance and compliance tests are usually rigidly specified even though they may not be well understood and may not correlate well with behaviour in service. Acceptance tests can probably be modified by mutual arrangement in the light of experience, but compliance tests are legally enforceable and are much more difficult to change.

Development tests are frequently *ad hoc* and may change as development proceeds. It is desirable to keep the tests as simple as possible so that behaviour can be understood and the effects of changes can be seen. Often it is not feasible to make full use of available information. For instance, it is difficult to estimate the strain rates in the plastic hinges of a thin walled structure which is crumpling on impact and, even if the variation in flow stress with strain rate is known for the material, this can probably be used only as a qualitative guide. Sometimes the object of interest cannot be tested directly and a surrogate specimen such as a scale model or a dummy person needs to be used. It may be difficult to relate the behaviour of these to that of the real object.

Research tests can be tailored to the project in hand but the test methods are often based on those used in the other two types of test. If impacts in service are common it should be possible to find correlations with behaviour in service, but if they are rare it is necessary to make a subjective assessment. Often this is little better than a guess.

Because of the greater number of variables and the frequent need to keep the performance of specific components and systems confidential there is less published discussion of test results than is found with impact testing of materials. Such published work as there is may be slanted towards publicity.

11.3.2 Non-destructive tests

A major feature of non-destructive impact testing is that more than one test can be made on the same specimen so that one source of variability is removed. It is also possible to test an item before it is put into service. There is not a clear-cut boundary between destructive and non-destructive testing and, in a given system, some components may be destroyed whilst others are not. Allowing for the destruction of subsidiary expendable components. Non-destructive testing of key components can be performed at any impact speed, but generally non-destructive testing is restricted to low speeds.

In the acceptance and development categories, typical non-destructive tests are for protection of the specimen in accidental impacts or for checking that it continues to function in a specified environment. Simple tests for accidental damage usually simulate accidental drops where the impact velocity is less than $5 \, \mathrm{m \, s^{-1}}$. There are a wide variety of such tests which are usually *ad hoc*. Typical ones are tumbling in a horizontal rotating drum, repeated raising and dropping by a revolving cam, or single drops. Measurement and analysis is often confined to a visual inspection with perhaps a record of peak acceleration. Tests of tuned anti-shock mountings are likely to include measurement of the impact pulse and shock spectra of the response.

Tests for functioning are of two main types. The first checks that the specimen continues to function after the impact is over. The second measures performance during the impact and is used, for instance, with accelerometers and load cells used in impact testing. Usually these are calibrated statically or under steady sinusoidal vibration at different frequencies and amplitudes and then spot checks are made to ensure that they continue to function during specified impacts.

Research tests often aim to establish the response of a specimen to a given impact pulse. They can cover a wide variety of test methods, measurement and analysis. When human volunteers are decelerated or accelerated, to establish lower bounds for the onset of injury, great care has to be taken with the test procedure but data retrieval and analysis is often very crude. When the natural frequencies and mode shapes of a structure are found by subjecting it to a measured blow, the test procedure may be very simple but the data retrieval and analysis are very sophisticated. Research tests are also used to validate computer models of systems or components. It can be difficult to obtain realistic data for initial input to the model and to measure in tests the parameters which are used in the model. Sometimes the computer model represents a surrogate specimen which can be tested and analysed more easily than the actual item of interest. The results may then be of doubtful value and there are a regrettable number of elaborate theoretical papers in which no attempt is made at experimental validation.

11.3.3 Destructive tests

The range of possible destructive tests on components or systems is very wide and, though some typical examples are discussed briefly, these are far from exhaustive. When the impact is produced by a collision between moving bodies there is a limit to the amount of kinetic energy which can be supplied, and this limits the size of the bodies used. This is shown in Fig. 11.6, which is drawn for a constant kinetic energy of $150 \, \mathrm{kJ}$ – approximately equivalent to a motor car of 1.5 tonnes weight travelling at $14 \, \mathrm{m \, s^{-1}}$. The moving weight can be very large at low velocities, but in hypervelocity tests it is only about $10^{-2} \, \mathrm{N}$. The line drawn for the performance of hypervelocity light gas guns

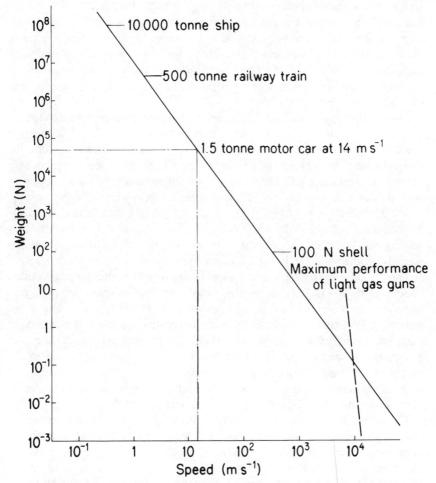

Figure 11.6 Reduction in moving mass with velocity at approximately 150 kJ energy.

indicates that the reduction in weight with speed is greater than this in practice. If the impact is produced directly by an explosion there is much less restriction on the energy input and the size of the impact is usually specified in a different way – by the weight of explosive used. Often this is given as an equivalent weight of a reference explosive such as TNT.

Impact tests on vehicles, vehicle components and injury preventing devices, such as safety helmets, are often at speeds below $14 \, \text{m s}^{-1}$ which can be obtained by dropping under gravity. Sometimes the specimen is pulled horizontally by a rope which passes over pulleys and is connected to a falling weight, and sometimes the weight is swung as a pendulum to give a horizontal impact. Complete vehicles can be run under their own power and using their own steering, or they can be propelled by an external source such as an electric motor and guided along a rail. As vehicle performance in accidental collisions improves it is important that the collection of accident data improves in parallel so that the effects of more sophisticated improvements can be seen (Macaulay and Penoyre, 1980).

In tests on cars and aircraft the main aim is to protect the vehicle occupants, and dummy occupants are regularly used in full size vehicles or sections of vehicles (Thomson *et al.*, 1984). In tests on ships the main aims are to limit the damage caused by the ship and to prevent the release of dangerous cargo. Ship occupants are seldom killed or injured by the impact itself. Tests are usually on scale models and frequently on dry land rather than in water (Hagiwara *et al.*, 1983). The effective mass of a ship is greater than its actual mass because of hydrodynamic effects in the water round the hull. Usually this is dealt with by adding on a fictitious virtual mass. Both dummy occupants and scale models present problems in the interpretation of results.

The response of large static structures is also usually tested on scale models (Booth *et al.*, 1983). It is sometimes simpler to analyse the response to an impulse which imparts an instantaneous velocity change to part of the structure. This can be simulated by a small explosive sheet which is separated from the model by a sheet of foam rubber to prevent surface damage. An explosive can be used which burns faster than the speed of sound in the specimen, so that the loading can be regarded as instantaneous. If the specimen is mounted on a fairly heavy pendulum the movement of this can be used to measure the applied impulse (Humphreys, 1965).

The response of structures to blast loading by a large explosion can be studied at full size or on scale models. There are two distinct aspects: the response of special protective structures to a defined blast, and the response of ordinary structures to varying amounts of blast (Norris *et al.*, 1959, Chapter 11). In terminal ballistics it is the interaction between a projectile and a target which is studied. The item of main interest can be either. In both of these areas testing appears to be largely empirical because there is appreciable scatter in test results and the processes involved are very complex and difficult to analyse.

In hypervelocity testing a major problem is attaining the appropriate impact speed. Tests are run in a vacuum, to eliminate air resistance, and accelerators fall broadly into two types: gun accelerators in which the moving body is guided as it accelerates, and explosive accelerators in which it is not. In gun accelerators the projectile is often held in a light disposable carrier to prevent it being worn by friction. In explosive accelerators the projectile is mounted directly on an explosive device (Kinslow, 1970, Chapter 1).

11.4 DATA RETRIEVAL

11.4.1 Residual deformations

The residual deformations remaining after the test is over can be measured statically. Quantitative information is usually confined to simple parameters such as the shortening of a thin walled tube or the volume of a crater. Qualitative information can also be obtained on the type of deformation, where it occurs and how it compares with deformation under static loading. Deformations of interest can range from the overall deflection of a large structure to microscopic changes in crystals and fibres (Zukas *et al.*, 1982, Chapter 4).

Relative velocities and rates of strain vary throughout the impact. Often this is ignored and maximum or average values are quoted, but sometimes repeat tests are made on a specimen, terminating each test when a predetermined amount of deformation has occurred and the relative velocity is still fairly close to its initial value. In this way the total deformation can be built up at an approximately constant velocity but with a substantially isothermal test. When the total deformation is reached in a single impact behaviour may be substantially adiabatic leading to different results (Vigness *et al.*, 1957).

11.4.2 Photography

(a) Use made of photography

Photographs record only displacements but they can do this in great detail. All types of impact can be recorded and the taking of photographs usually has no effect on the test itself. A major use of photographs is to allow impacts to be studied qualitatively at leisure and this can give considerable insights into patterns of behaviour without further analysis.

It can be difficult to obtain quantitative information from photographs and usually it is feasible to make only a small number of measurements. The locations of points of interests are measured on successive photographs to give a digital record of how they move with time. Usually the process involves a human operator and it can be lengthy. Velocities obtained by differentiating displacements are likely to be usable but accelerations obtained by double differentiation are likely to be suspect.

Sometimes clouds of debris are produced during an impact. These obscure behaviour of interest unless X-ray high speed photography is used. X-ray photography can also be used to a limited extent to record what is happening inside a test specimen.

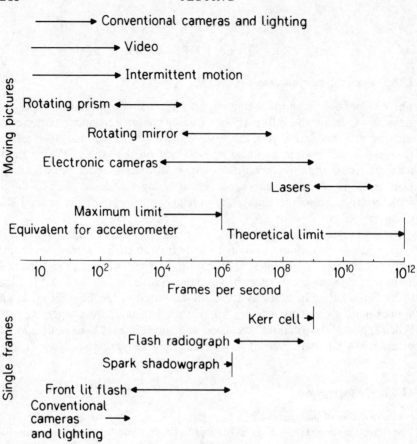

Figure 11.7 Camera speeds.

(b) Cameras

The range of framing times for high speed cameras is shown in Fig. 11.7 (Zukas *et al.*, 1982, Chapter 7, Figs 3, 10). With a still camera the framing time is the same as the exposure time, but with a cine camera the framing time is the time between successive exposures and the exposure time is typically only about 20% of this. High speed pictures are almost invariably recorded on photographic film although the camera optical system may include some electronic image enhancement. Photographic film is a parallel recording medium with all the data on each frame recorded simultaneously. Electronic video recorders are series recording devices, with the data in each frame being built up sequentially by a scanning spot, and they are too slow for most impact tests.

In general the quality of the picture decreases as the framing time decreases,

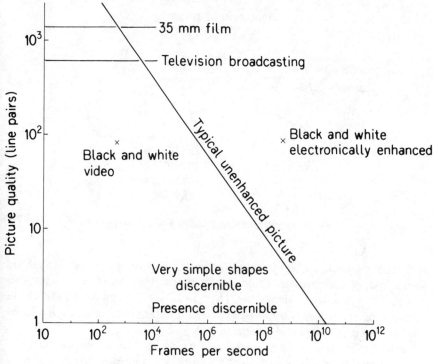

Figure 11.8 Deterioration in picture quality with exposure rate.

broadly as shown in Fig. 11.8. The quality of the picture is expressed in line pairs. A good 35 mm still photograph is equivalent to 1200 line pairs and a broadcast television picture is about 600. At three line pairs very simple shapes can be seen roughly, and at one only the presence or absence of an object can be detected. The behaviour shown in Fig. 11.8 can be modified by image enhancement (Zukas *et al.*, 1982, Chapter 7, Fig. 1).

In conventional cine cameras the film moves intermittently. It is stationary during the exposure and moving rapidly between exposures. The upper speed limit is determined by the strength of the film and the wear of the intermittent feed mechanism. The framing speed can be increased appreciably if the film moves continuously and the image is made to move at the same speed by a rotating prism, as shown in Fig. 11.9. Picture quality is not so good as with an intermittent motion camera because the motion compensation is not perfect. The upper speed limit is limited by the strength of the film and the centrifugal stresses in the prism. The films are the same as those used in intermittent motion cameras and can be shown on conventional cine projectors (Zukas *et al.*, 1982, Chapter 7, Fig. 11).

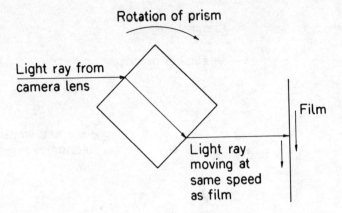

Figure 11.9 Principle of rotating prism camera.

Framing speed can be increased further by placing a strip of film on the inside wall of a continuously rotating drum and replacing the heavy rotating prism by a lighter rotating mirror. The film length is restricted and there are limitations on the lenses which can be used. The film is usually unsuitable for showing in a conventional cine projector.

A further increase in speed is obtained by placing a strip of film on the inside wall of a stationary drum, and situating a number of lenses in an arc in front of the film. This gives effectively a set of individual cameras. The image of the test event is focused on to a mirror which rotates continuously about the central axis of the drum and reflects into each camera in turn. The number of pictures is limited, synchronizing the camera with the test is difficult, and the lights need to be extinguished or the shutter closed before the film is exposed a second time. Explosively activated shutters are often used.

As well as cine films, still photographs can be taken. There is no clear transition; the stationary drum camera can be regarded either as taking a moving picture or a set of still ones, whilst banks of still cameras can be used to cover a complete impact. In still cameras the exposure is obtained either by having the shutter open continuously and using a short duration flash or by having continuous illumination and using a fast acting shutter. Short duration flashes can be obtained from electric sparks or pulsed lasers. High intensity X-ray flashes can also be produced. Very high speed shutters can be provided by a number of transparent materials which polarize light when they are subjected to high electrostatic or magnetic fields. In combination with conventional polarizing materials they can be arranged so that light is transmitted only when the field is applied. The best known of these devices is the Kerr cell (Zukas *et al.*, 1982, pp. 249–51).

Usually it is necessary to have very good illumination because exposure times are so short. If colour films are wanted, all of the light reaching the film needs to come from this illumination. If black and white pictures are acceptable, electronic image intensification in the optical system of the camera can increase the light reaching the film by up to 10^5. This needs to be a parallel intensification system covering the whole frame simultaneously so that the parallel recording capability of the film is used (Zukas *et al.*, 1982, pp. 251–5).

A special application of photography is the use of a smear or streak camera. In this the film moves continuously past a narrow slit giving a continuous record of the movement of an object along or across the slit. The result is akin to the output from an electronic transducer recording displacement against time, and there is no clear distinction between a streak camera and electronic instrumentation using optical transducers. Another use of a streak camera is to provide a timing device. For instance, an argon filled groove covered by a transparent plastic block will glow when a high intensity shock wave reaches it. The start of the luminous streak on a continuously moving film gives the time at which the shock wave arrives.

(c) Analysis

A great deal can be learnt from visual inspection of photographs without any measurements being taken. It is common practice to coat different items with non-reflecting paint of different colours so that their relative movement can be seen more easily. If measurements are to be made it is necessary to select some reference points, and it can be useful to mark these in a distinctive way such as with the butterfly shown in Fig. 11.10. The point where the triangles meet can be located fairly accurately, even on a poor quality photograph, and the circle appears as an ellipse. From the ratio and angle of the major and minor axis the plane in which the marker lies can be estimated. It is useful to have a stationary

Figure 11.10 Butterfly marker.

reference point as well as moving ones because successive frames do not line up precisely.

Usually the framing speed obtained by setting the camera controls is only approximate and it is useful to have an accurate timing device visible in each frame. This can be a flashing diode incorporated in the camera, or a separate device such as a rotating pointer or drum placed in the field of view. Flashes can be triggered by specific events, such as initial contact, to give accurate cross-checks between films and electronic instrumentation.

It is desirable to have more than one camera. These give different viewpoints and act as back-ups in case one camera fails. Often it is useful to have a simple conventional cine camera as well to take an overall view of the test. This gives overall impressions and can also provide useful information when a test goes wrong outside the impact area.

Analysing devices are available in which films can be back projected, one frame at a time, on to a ground glass screen. An operator lines up cross-wires on each reference point in turn and its co-ordinates are recorded automatically. The process can become very tedious and results are subject to background noise from operator error. Only displacements are obtained and these usually need to be smoothed before they are used. Usually simple averaging methods are used and the amount of smoothing is decided on intuitively. Computer programs can be written to give various forms of analysis but the usual limitations are encountered if the displacements are differentiated to give velocities and accelerations.

11.4.3 Electronic instrumentation

(a) Use made of electronic instrumentation

The main advantage of electronic instrumentation is that it can provide continuous records in a form suitable for automatic processing. It can also give accelerations and velocities directly, eliminating the problems of differentiating empirical results. The disadvantages are that a separate channel is needed for each item recorded, that transducers are easily damaged and that they can modify the behaviour of the test specimen. Records are often very noisy and spurious signals, produced for instance by leads in violent motion, may be indistinguishable from real signals.

The range of frequencies for some different types of transducer are shown in Fig. 11.11 (Trade and Technical Press, 1972, p. 114). It is not possible to make a direct comparison with the framing rates of high speed cameras but, assuming that a sinusoidal signal is being recorded, and that ten points are needed to define a complete cycle of a sine wave, a camera speed of 10 frames per second is broadly equivalent to a frequency response of 1 Hz. On this basis electronic instrumentation can be used up to the equivalent of about 10^6 frames per second compared with, say, 10^{10} frames per second for photography.

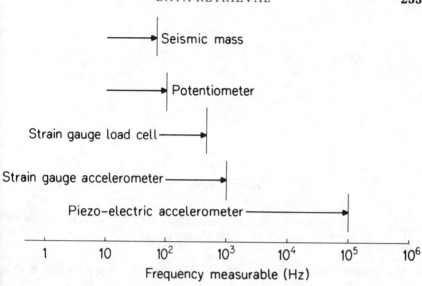

Figure 11.11 Electronic transducer speeds.

In general, electronic instrumentation is used extensively in lower speed impacts, to a limited extent at ballistic velocities and very little in hypervelocity impacts. This is partly because of the limits on response rates and partly because no transducer can withstand the forces which occur in very high speed impacts.

(b) Accelerometers

Basically an accelerometer is a spring–mass system with one degree of freedom. The spring is attached to a base plate which is attached to the location at which acceleration is to be measured, as shown in Fig. 11.12. If an acceleration is applied to the base plate the equation of motion of the mass is

$$m(a - \ddot{x}_r) = Kx_r$$

where m is the mass, K the spring stiffness, a the applied acceleration and x_r the displacement of the mass relative to the base plate. This gives

$$x_r = \frac{m}{K}(a - \ddot{x}_r) \tag{11.9}$$

If a constant acceleration is applied slowly, $\ddot{x}_r \rightarrow 0$ and $x_r = ma/K$, giving a constant relationship between x_r and a. As long as the time taken for the acceleration to change measurably is more than about five times the natural period of the spring–mass system, this constant relationship is maintained.

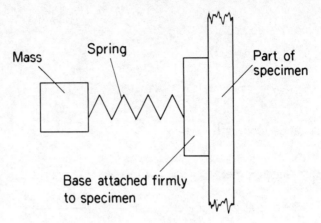

Figure 11.12 Simple accelerometer.

If a constant acceleration is applied instantaneously it can be shown that

$$x_r = \frac{ma}{K}(1 - \cos \omega t) \qquad (11.10)$$

where ω is the natural frequency of the spring–mass system, when the base plate is held rigidly. This represents a vibration of the mass with a peak amplitude of $x_r = \pm ma/K$ about a mean value of $x_r = ma/K$. Apart from the value of the constant term, this equation is identical to Equation 6.16, which was derived for a constant force applied instantaneously to the mass (Open University, 1974, pp. 20–8). This applies in general to all the discussion in Chapter 6 on the response of simple vibrating systems, and so the value of x_r depends on the shape and duration of the applied acceleration pulse as well as on its amplitude. As the accelerometer is being used to measure the applied acceleration it is difficult or impossible to disentangle those effects, and often all that can be said is that there are likely to be vibrations of significant but unpredictable amplitude at the natural frequency of the accelerometer. These can completely mask the slower part of the response in which x_r is linearly proportional to a, and need to be removed either by damping or by filtering.

In addition there is likely to be some form of resonant vibration excited in the item which the accelerometer is attached to, and this usually needs to be filtered out before the underlying shape of the acceleration pulse can be seen. A typical unfiltered record from an accelerometer is shown in Fig. 11.13(a). This can be smoothed by eye or by digital techniques, or it can be filtered to remove the higher frequencies, to give the approximate acceleration pulse shown in Fig. 11.13(b). There is no correct method which gives a definitive version of the initial pulse, and so the method and values used should be specified (Searl et al., 1971).

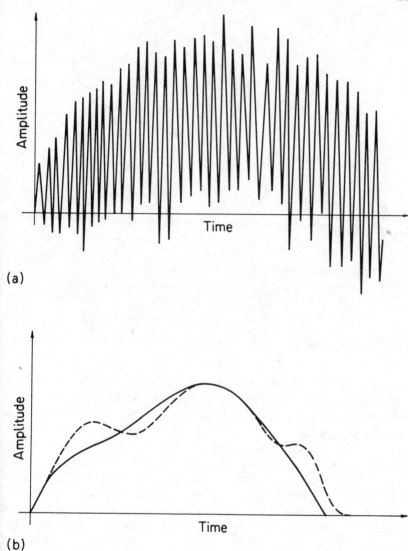

Figure 11.13 Accelerometer records. (a) Unprocessed trace. (b) Approximations to underlying pulse.

As well as leading to problems with analysing records, the superimposed vibrations make it difficult to calibrate accelerometers under impact loading. They also reduce the sensitivity because the accelerometer has to withstand the maximum amplitude of vibration without damage.

Usually calibration is done with steady-state sinusoidal inputs of

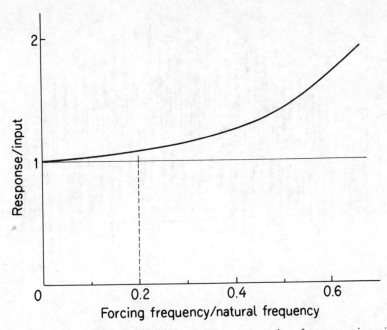

Figure 11.14 Response of simple spring–mass system to low frequency sinusoidal excitation.

different frequencies and amplitudes followed by some simple bump tests to check that the accelerometer continues to function under shock loading. The lower frequency range when a sinusoidal input of constant peak acceleration is applied to the base of an accelerometer is plotted in Fig. 11.14. It can be seen that for an undamped accelerometer there is a substantially constant relationship between the peak acceleration applied to the base and the peak value of x_r up to about 20% of the natural frequency. With optimally damped accelerometers the substantially constant relationship extends to about 1.5 times the natural frequency, but at the expense of the output lagging behind the input in a way which varies non-linearly with frequency (Open University 1974, pp. 28–44).

The most widely used accelerometers have a piezo-electric crystal as the spring. This is relatively very stiff; thus the natural frequency can be very high but the spring deflection is far too small to measure directly. The electrical charge on the crystal varies linearly with the deflection but is much more sensitive, and it is this which is measured (Open University, 1974, pp. 45–9).

(c) Other transducers

Change in acceleration is an absolute parameter and requires no fixed reference point. Changes in velocity or displacement need to be measured from a datum, which can be provided either by an external fixed point or by a seismic mass. Because accelerometers are widely available, velocity and acceleration are often obtained by integrating an accelerometer signal and selecting a suitable stationary reference point to give an appropriate constant of integration.

Velocity can be measured more directly by having two or more fixed contacts a known distance apart and recording the time taken by a moving contact to move from one to the next. It can also be measured by moving a magnet relative to a coil, when a current is produced in the coil which is directly proportional to their relative velocity, or by measuring the Doppler frequency shift in a modulated radar or laser beam.

Displacement can be measured by a potentiometer in which the resistance is proportional to the location of a moving contact. For fairly small displacements a linear potentiometer can be used, and for larger ones a rotary potentiometer with a string wound several times round a pulley or shaft. Displacement in three dimensions can be measured by combinations of linear and rotary potentiometers. Very small displacements can be measured by the change in capacitance between a fixed and a moving plate.

If a seismic mass is used as a datum the transducer can be similar to the accelerometer shown in Fig. 11.12. If the system has a relatively low natural frequency, so that the impact lasts less than 20% of the natural period, and if the system is undamped, the mass does not have time to respond to the movement of the base. The mass can then be regarded as stationary during the impact, though it can vibrate violently afterwards (Trade and Technical Press, 1972, pp. 110–13).

Forces can be measured by load cells. These can use resistance strain gauges or piezo-electric crystals. Load cells usually vibrate in the same way as accelerometers and the signals need to be smoothed in the same way. Because a load cell usually needs to be appreciably larger than an accelerometer it has a much lower resonant frequency, and so the maximum frequency which can be measured is also lower.

(d) Signal processing

A typical signal processing system is shown in Fig. 11.15. The electrical signal from the transducer is amplified and stored. It may be analysed fully or partly before being stored. Analysis before storage needs to be done in real time during the test but analysis after storage can be done with the signal replayed at any convenient speed. If the signal is weak and needs to be transmitted more than a very short distance the amplification is likely to be in two stages, with a pre-amplifier at the start of the transmission line and the main amplifier at the end. The transmission line can be a conducting wire, a telemetry system or an optical link.

The signal processing system has its own dynamic response and can

Transducer pre-amplifier

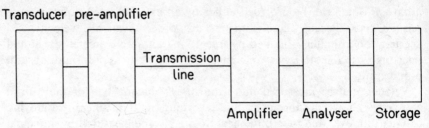

Figure 11.15 Simple signal processing system.

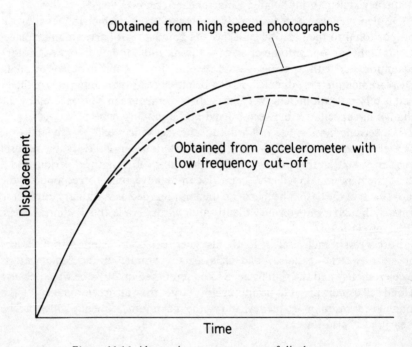

Figure 11.16 Alternative measurements of displacement.

attenuate or amplify a signal pulse or produce resonant vibrations in broadly the same way as a mechanical vibrating system does. It can also generate background noise, some of which is continuous and some transient and indistinguishable from the real signal. This noise originates within the equipment and can also be produced, for instance, by the violent movement of leads during the impact. The signal/noise ratio imposes a limit on the smallest signal which can be processed.

Up to the end of the amplification stage the signal is generally in analogue form with the amplitude of the electrical signal being proportional to the amplitude of the parameter being measured. After amplification the signal can be digitized so that further processing can be analogue, digital or a mixture of the two. Digital processing is inherently more accurate and flexible but analogue processing may be more convenient.

Usually amplifiers provide substantially constant amplification of the input signal over an operating range of frequencies and attenuate the signal at frequencies above and below this range. For a given nominal upper frequency limit the effective upper limit varies with the information needed because phase relationships among different frequency components in the signal become distorted at frequencies below those at which attenuation occurs. Frequency limits introduced by amplifiers or by other components of the signal processing system may be more restrictive than those imposed by the transducer itself.

The lower frequency limit can be reduced to zero so that constant amplitude signals are amplified, but this introduces problems with reference values drifting away from their original settings and complicates the setting-up procedure. A finite lower frequency limit eliminates these problems but attenuates the low frequency components of a signal. This is shown for a displacement obtained by integrating an accelerometer signal in Fig. 11.16. Often this is relatively unimportant, or can be allowed for, and the greater ease of use with a low frequency cut-off is more important.

(e) Analysis

Signals in an electrical form, whether analogue or digital, can be analysed readily with a minimum of attention by the operator. There are numerous standard analysis devices and computer programs, and usually it is advisable to use a known and tried system even if it is not the ideal method for the analysis in hand. For instance, digitized signals are often routinely converted into their Fourier transforms because digital Fourier analysis methods are well developed, widely available and give results in a known format.

Most analysis methods lose details of the signal and introduce distortions such as time delays, phase shifts and variations in scaling factors with frequency. Specification of acceptable distortion is usually subjective and it is advisable to report the assumptions made and the conventions used. Often the presence of distortions is not appreciated and results can be misinterpreted.

Some analysis can take place anywhere along the signal processing system in a way which depends on the equipment available and the information needed. Amplification is analogue; filtering, integration and differentiation, multiplication and division can be analogue or digital; and more advanced analysis is usually digital. There are few clear-cut boundaries and a procedure performed as a final digital analysis in one set of circumstances may be performed as routine analogue signal processing prior to analysis in another.

Bibliography

The literature on impact is extensive but widely dispersed, unevenly distributed among different topics and on many levels of complexity. The literature on non-impact behaviour is more coherent and includes an appreciable number of relatively simple textbooks at about undergraduate level.

No attempt is made in the Bibliography and References to be comprehensive, to give direct reference to original sources or to keep abreast of the state of the art. Reference is usually made to secondary sources such as textbooks or monographs which give an overview of their subject areas and usually include extensive bibliographies leading back to the original sources. Papers, monographs or conference proceedings are usually given as amongst the main references available on a subject, but where textbooks are quoted, the choice is largely subjective and other textbooks may be equally good and better suited to a particular reader.

Often in quoting published results I have modified them, in an attempt to produce a coherent picture at a uniform level of difficulty. The underlying results have not been altered, but at first glance my presentation may sometimes appear unlike that in the reference quoted.

The journals *Shock and Vibration Digest* and *International Journal of Impact Engineering* regularly carry papers dealing with impact and related subjects. These cover the latest developments but are usually for a specialist audience and at a technical level appreciably above that of this book. They should be consulted for information on the state of the art, but papers also appear largely at random in a wide range of other journals.

GENERAL SOURCES

Three books are particularly useful as overall sources of information on impact. All three are based on courses of lectures and, as they were published at ten-year intervals, they give a reasonably good overall view of the subject and its development. In chronological order they are:

Goldsmith, W. (1960) *Impact*, Edward Arnold.
This is intended for physicists and applied mathematicians and can be difficult in places. It gives a résumé of work published before the advent of modern instrumentation and computing.

Johnson, W. (1972) *Impact Strength of Materials*, Edward Arnold.
This is based on a lecture course for engineering M.Sc. students and is generally easier and less rigorous than Goldsmith. It contains a collection of photographs covering a wide range of topics.

Zukas, J.A. *et al.* (1982) *Impact Dynamics*, Wiley.
This is intended for practising engineers, particularly those in United States military and space projects. It includes surveys of recent instrumentation and computing.

CHAPTERS 1–4

General

A useful textbook covering the non-impact side of all four chapters is
McClintock, F.A. and Argon, A.S. (eds) (1966) *Mechanical Behaviour of Materials*, Addison-Wesley.
This is intended for engineering undergraduates but also contains a large amount of additional material at a more advanced level.

There does not appear to be an equivalent book on impact behaviour.

Chapter 1

Static three-dimensional analysis of small deflections in a linearly elastic material is dealt with in
Ford, H. with Alexander, J.M. (1977) *Advanced Mechanics of Materials*, 2nd edn, Ellis Norwood.
The analysis is worked out in full as well as in condensed notation.

An alternative presentation in condensed notation is given in
Freudenhal, A.M. (1966) *Introduction to the Mechanics of Solids*, Wiley.

The classical monograph on stress waves is
Kolsky, H. (1963) *Stress Waves in Solids*, Dover.
This aims to present classical theory at a level which is easy for physics graduates.

A comprehensive but difficult book is
Achenbach, J.D. (1973) *Wave Propagation in Elastic Solids*, North Holland/American Elsevier.

Chapter 2

The behaviour of rubber-like materials is dealt with in
Freakley, P.K. and Payne, A.R. (1978) *Theory and Practice of Engineering with Rubber*, Applied Science Publishers.
Williams, D.J. (1971) *Polymer Science and Engineering*, Prentice-Hall.

Flügge, W. (1967) *Viscoelasticity*, Blaisdell.
This is intended as an introductory book for self-study. It is purely theoretical and gives no specific applications but covers the subject in a simple basic way.

Rouse, H. and Howe, J.W. (1953) *Basic Mechanics of Fluids*, Wiley.
This is intended for average undergraduates.

Kinslow, R. (ed.) (1970) *High Velocity Impact Phenomena*, Academic Press.
This has a section on shock waves in solids. It is intended for scientists and engineers not already familiar with the subject.

Nowacki, W.K. (1978) *Stress Waves in Non-Elastic Solids*, Pergamon.
This is a comprehensive but closely packed and difficult monograph.

Kolsky, H. and Prager, W. (1964) *Stress Waves in Inelastic Solids*, Springer Verlag.
This contains a number of interesting papers.

Chapter 3

Non-impact aspects of plasticity are dealt with in McClintock and Argon (1966) and Ford and Alexander (1977) (see above), and in
Calladine, C.R. (1969) *Engineering Plasticity*, Pergamon.
This is intended for engineering undergraduates.

Impact aspects are dealt with in Johnson (1972) (see above), and in
Campbell, J.D. (1970) *Dynamic Plasticity of Metals*, Springer Verlag.
This is closely packed and intended for physicists. It can be difficult in places, but is an excellent starting point for further study.

Chapter 4

Useful fairly simple textbooks on static fracture mechanics are
Knott, J.F. (1973) *Fundamentals of Fracture Mechanics*, Butterworth.
This is written by a metallurgist who tries to link the metallurgical and engineering approaches to the subject.
Lawn, B.R. and Wilshire, T.R. (1975) *Fracture of Brittle Solids*, Cambridge.
This covers macroscopic, microscopic and dislocation behaviour.

Low temperature transitions from ductile to brittle fracture are dealt with in
Wigley, D.A. (1971) *Mechanical Properties of Materials at Low Temperatures*, Plenum Press.

There appear to be no equivalent books on impact behaviour, but useful volumes of conference proceedings are
Impact Testing of Metals, ASTM Special Technical Publication 466, 1969.
Dynamic Crack Propagation (ed. G.C. Sih), Noordhoff International Publishing, 1973.
Dynamic Fracture Toughness (conference technical director M.G. Dawes), The Welding Institute, 1977.
Advances in Fracture Research, Proceedings of 5th International Conference on Fracture (ed. D. Francois), Pergamon, 1981, vols 1 and 2.

CHAPTERS 5–7

General

Basic textbooks on dynamics for engineers tend to concentrate on differential equations. More advanced engineering textbooks introduce Lagrangian dynamics, but to fine a simple introduction to integral equations it is necessary to turn to basic textbooks for physicists. Two useful books are:

Troup, G.J. (1975) *Mechanics*, Longman.
This is a simple introductory text for first-year science undergraduates.

Kibble, T.W.B. *Classical Mechanics*, McGraw-Hill.
This is an introductory text for physics undergraduates.

A more advanced textbook for engineers is

Hurty, W.C. and Rubinstein, M.F. (1964) *Dynamics of Structures*, Prentice-Hall.
This deals with structural dynamics but emphasizes methods of analysis rather than applications. It is intended for final-year undergraduates or first-year postgraduates and includes differential equations, integral equations, energy methods and computing.

Chapter 5

The impact of rigid bodies is discussed in Goldsmith (1960, Chapter 2) (see above).

For a comprehensive introduction to Lagrangian dynamics, see

Walls, D.A. (1967) *Lagrangian Dynamics*, McGraw-Hill.

Chapter 6

The following textbooks include response to transient inputs and pulse inputs as well as steady-state vibrations:

Jacobson, L.S. and Ayre, R.S. (1958) *Engineering Vibrations*, McGraw-Hill.
This includes a large number of response spectra and cites references containing many more.

Morrow, C.T. (1963) *Shock and Vibration Engineering*, Wiley.

Biggs, J.M. (1964) *Introduction to Structural Dynamics*, McGraw-Hill; Chapter 2.

Broch, J.T. *et al.* (1980) *Mechanical Vibration and Shock Measurements*, Bruel and Kjaer.

Chapter 7

Snowdon, J.C. (1968) *Vibration and Shock in Damped Mechanical Systems*, Wiley.
This deals with the response of damped vibrating systems.

Elastic/ideally-plastic systems are dealt with in Biggs (1964) (see above).

Introductory texts on Laplace transforms are:

Sensicle, A. (1968) *Introduction to Control Theory for Engineers*, Blackie.

Spiegel, M.R. (1965) *Laplace Transforms*, Schaum.

Stress waves are dealt with in Johnson (1972) and Kolsky (1966) (see above).

CHAPTERS 8–10

Chapter 8

The design of energy absorbers is largely empirical and there does not seem to be an overall approach covering the entire range of impact speeds. Various individual aspects have been studied extensively but largely on an *ad hoc* basis; a topic can become the subject of considerable activity for a time and then be neglected. Some absorbers are commercially available and some methods are patented.

For low impact speeds and accelerations up to, say, $10\,g$, rubber is often used as an energy absorber, and books such as Freakley and Payne (1978) and Snowdon (1968) (see above) may be consulted.

At speeds up to, say, $30\,\mathrm{m\,s^{-1}}$ permanent deformation of metallic energy absorbers is common, and useful review articles are:
Ezra, A.A. and Fay, R.J. (1972) An assessment of energy absorbing devices for prospective use in aircraft impact situations, in *Dynamic Response of Structures* (eds G. Herrmann and N. Perrone), Pergamon.
Johnson, W. and Reid, S.R. (1978) Metallic energy dissipating systems. *Applied Mechanics Reviews*, **31**(3).

Penetration and perforation at ballistic velocities have been studied extensively since the early nineteenth century. A comprehensive review article is:
Backman, H.E. and Goldsmith, W. (1978) The mechanics of penetration of projectiles and targets. *International Journal of Engineering Science*, **16**, 1–99.

Metal targets are dealt with in Zukas *et al.* (1982, 'Penetration and perforation of solids') (sea above).

Kornhauser, M. (1967) *Structural Effects of Impact*, Cleaver-Hume, Chapter 3.
This deals with soil, rock and concrete.

Penetration of concrete at speeds up to about $60\,\mathrm{m\,s^{-1}}$ is discussed in
Kennedy, R.P. (1976) A review of the procedures for the analysis and design of concrete structures to resist missile effects. *Nuclear Engineering and Design* (37).
Barr, P. (1983) Studies of the effects of missile impacts on structures, *Atom* (318).

Energy absorption at hypervelocities is dealt with in Backman and Goldsmith (1978) and Zukas *et al.* (1982) (see above), and in
Gehring, J.W. Theory of impact on thin targets and shields and correlation with experiment, in Kinslow (1970) (see above).
Gehring, J.W. Engineering considerations in hypervelocity impact, in Kinslow (1970) (see above).

Chapter 9

Useful textbooks on the dynamic response of structures are Hurty and Rubinstein (1964) (see above) and:

Norris, C.H. *et al.* (1959) *Structural Design for Dynamic Loads*, McGraw-Hill.
This is intended as a review of the subject for structural engineers principally interested in civil engineering structures.

Craig, R.R. (1981) *Structural Dynamics: An Introduction to Computer Methods*, Wiley.
This is intended as background information for engineers using structural dynamics computer programs.

Impulsive loading of beams is dealt with in Johnson (1972, Chapter 6) (see above), and there are a number of interesting papers in

Noor, A.K. and McComb, H.G. (1978) *Trends in Computerized Structural Analysis and Synthesis*, Pergamon.

More specific applications of computer analysis are contained in Zukas *et al.* (1982, Chapters 10 and 11) (see above), and

Davies, G.A. (1984) *Structural Impact and Crashworthiness*, Elsevier: vol. 1, Chapter 7.

There are also commercially available computer programs and specialist computing bureaux.

Chapter 10

The literature on impact injury is extensive, widely dispersed and often controversial. It ranges from abstract mathematical papers to clinical descriptions of specific injuries.

A short overall review article is
Von Gierke, H.E. (1964) Biodynamic response of the human body. *Applied Mechanics Reviews*, **17**, 951–8.

A much longer review article is
Snyder, R.G. (1970) Human impact tolerance. *International Automobile Safety Conference Compendium*, SAE, 712–55.

Current reviews which are aimed at motor car crashes are
Human Tolerance to Impact Conditions as Related to Motor Vehicle Design, SAE, 1980.
Mathematical Simulation of Occupant and Vehicle Kinematics, SAE, 1984.

Three series of conference proceedings cover current research on impact injury. In these, topics appear more or less at random and there are wide variations in points of view, so the non-specialist can easily become confused. They are
Stapp Car Crash Conference Proceedings, SAE (annual).
Biomechanics of Impact Proceedings, IRCOBI (annual).
International Technical Conferences on Experimental Safety Vehicles, NHTSA, US Department of Transportation (biennial).

The standard work on the static stress–strain properties of various components of the human body is
Yamada, H. (1970) *Strength of Biological Materials*, Williams & Wilkins, Boston.

CHAPTER 11

Much impact testing is *ad hoc*, directed at a specific type of impact, and confidential for either commercial or military reasons. Much of the expertise involved arises from extensive experience and has not been analysed and codified. There does not seem to be an overall approach covering the whole range of impacts.

Equipment for collecting data from tests is usually available commercially and the manufacturers provide both literature and advice. Generally this is excellent but it tends to be rather optimistic, to gloss over difficulties and to concentrate on one way of doing things.

Two excellent reviews in Zukas *et al.* (1982) (see above) deal with material testing and high speed photography. These are Chapter 8, 'Material behaviour at high rates of strain', and Chapter 7, 'Image forming instruments'.

There does not seem to be a similar review article on electronic instrumentation, but Broch *et al.* (1980) (see above) contains useful information.

References

Abrahamson, G.R. and Goodier, J.N. (1966) Dynamic flexural buckling of rods within an axial plastic compression wave. *Journal of Applied Mechanics*, June, 241–7.

Abrahamson, G.R. and Lindberg, H.E. (1972) Peak load characterization of critical pulse loads in structural dynamics. In *Dynamic Response of Structures* (eds G. Herrmann and N. Perrone), Pergamon.

Aldman, B. *et al.* (1982) Introcranial deformation patterns due to impulsive loading – a model study, 9th International Technical Conference on Experimental Safety Vehicles, NHTSA.

al Hassani, S.T.S *et al.* (1972) Characteristics of inversion tubes under axial loading. *Journal of Mechanical Engineering Science*, **14**, 370.

American Society of Metals (1975) *Metals Handbook*, 8th edn. Vol. 10, *Toughness and Fracture Mechanics*, Figs 11 and 12.

Ammon, W. (1984) Applicability of dynamic plasticity theorems to impulsively loaded reinforced concrete structures. In *Structural Impact and Crashworthiness*, Vol. 2 (ed. J. Morton), Elsevier.

Anderson, W.F. *et al.* (1983) Optimisation of rock/polymer composites to resist projectile penetration. *Matériaux et Constructions*, **16**(95), 343.

Andrews, K.P.F. *et al.* (1983) Classification of the axial collapse of cylindrical tubes under quasi-static loading. *International Journal of Mechanical Sciences*, **25**, 687–96.

Andrzejewski, A. *et al.* (1981) Experimental determination of high loading rate effects on fracture toughness of aluminium alloys. In *Advances in Fracture Research*, Proceedings 5th International Conference on Fracture (ed. D. Francois), Pergamon.

Barber, N.F. (1961) *Experimental Correlograms and Fourier Transforms*, Pergamon.

Begeman, P.C. *et al.* (1973) Spinal loads resulting from -Gx acceleration. In 17th Stapp Conference.

Biggs, J.M. (1959) Simplified analysis and design for dynamic load. In *Structural Design for Dynamic Loads* (C.H. Norris *et al.*), McGraw-Hill.

Biggs, J.M. (1964) *Introduction to Structural Dynamics*, McGraw-Hill.

Bodner, S.R. and Symonds, P.S. (1972) Experimental and theoretical investigation of the plastic deformation of cantilever beams subjected to impulsive loading. *Journal of Applied Mechanics*, December, 719–28.

Booth, E. *et al.* (1983) Impact scalability of plated steel structures. In *Structural Crashworthiness* (eds N. Jones and T. Wierzbicki), Butterworth.

Brabin, E.J. (1968) Energy absorbing devices used in crash testing. Motor Industry Research Association Bulletin no. 2.

Broch, J.T. *et al.* (1980) *Mechanical Vibration and Shock Measurements*, Bruel and Kjaer.

Calder, C.A. and Goldsmith, W. (1971) Plastic deformation and perforation of thin plates resulting from projectile impact. *International Journal of Solids and Structures*, **7**, 863.

Calladine, C.R. (1969) *Engineering Plasticity*, Pergamon.

Campbell, J.D. (1970) *Dynamic Plasticity of Metals*, Springer Verlag.

Cantwell, W. (1984) A study of the impact resistance and subsequent O-compression fatigue performance of non-woven and mixed woven composites. In *Structural Impact and Crashworthiness*, Vol. 2 (ed. J. Morton), Elsevier.

Chandler, R.F. and Christian, R.A. (1969) Comparative evaluation of dummy performance under -*Gx* impact. In 13th Stapp Conference.

Christopher, P.R. *et al.* (1981) An interlaboratory programme of dynamic fracture toughness tests. In *Advances in Fracture Research*, Proceedings of 5th International Conference on Fracture (ed. D. Francois), Pergamon, Vols 1 and 2.

Congleton, J. (1973) Practical applications of crack-branching measurements. In *Dynamic Crack Propagation* (ed. G.C. Sih), Noordhoff.

Conway, M.D. and Jakubowski, M. (1969) Axial impact of short cylindrical bars. *Journal of Applied Mechanics*, **36**, 809.

Cottrell, A.H. (1957) Deformation of solids at high rates of strain. In *The Properties of Materials at High Rates of Strain*, Institution of Mechanical Engineers.

Currey, J.D. (1979) Changes in the impact energy absorption of bone with age. *Biomechanics*, **12**, 459–69.

Damon, A., Stoudt, H.W. and McFarland, R.A. (1966) *The Human Body in Equipment Design*, Harvard.

Davidson, J.F. (1953) Buckling of struts under dynamic loading. *Journal of Mechanics and Physics of Solids*, **2**, 433–40.

Davies, R.M. (1956) *Stress Waves in Solids*, Cambridge.

de Rouvray, A. *et al.* (1984) Numerical techniques and experimental validations for industrial applications. In *Structural Impact and Crashworthiness*, Vol. 1 (ed. G.A. Davies), Elsevier.

Eftis, J. and Krafft, J.M. (1965) A comparison of the initiation of the rapid propagation of a crack in a mild steel plate. *Journal of Basic Engineering (ASME Transactions D)*, **87**, 257–63.

Evensen, H.A. and Evan-Iwanowski, R.M. (1966) Effects of longitudinal inertia upon the parametric response of elastic columns. *Journal of Applied Mechanics*, **33**, 141–8.

Fan, W.R.S. (1971) Internal head injury assessment. In 15th Stapp Conference.

Fearnhough, G.D. (1973) The small-scale test and its application to fracture propagation problems. In *Dynamic Crack Propagation* (ed. G.C. Sih), Noordhoff.

Florence, A.L. and Firth, R.D. (1965) Rigid-plastic beams under uniformly distributed impulses. *Journal of Applied Mechanics*, September, 481–8.

Flügge, W. (1967) *Viscoelasticity*, Blaisdell.

Ford, H. with Alexander, J.M. (1977) *Advanced Mechanics of Materials*, 2nd edn, Ellis Norwood.

Forrestal, M.J. *et al.* (1980) An explosive loading technique for the uniform expansion of 304 stainless steel cylinders at high strain rates. *Journal of Applied Mechanics*, **47**, 17.

Fowler, J.E. and Newman, K.F. (1980) The use of computer simulation for the design of safer vehicles. In *Progress Towards Safer Passenger Cars in the United Kingdom*, Institution of Mechanical Engineers.

Freakley, P.K. and Payne, A.R. (1978) *Theory and Practice of Engineering with Rubber*, Applied Science Publishers.

Freudenhal, A.M. (1966) *Introduction to the Mechanics of Solids*, Wiley.

Glaister, D.H. (1978) Human tolerance to impact acceleration. *Injury*, 9.

Gögler, E. *et al.* (1977) Biomechanical experiments with animals on abdominal tolerance levels. In 21st Stapp Conference.

Goldsmith, W. (1960) *Impact*, Edward Arnold.

Goldsmith, W. (1981) Current controversies in the stipulation of head impact criteria. Letter to the Editor, *Journal of Biomechanics*, **14**, 883–4.

Grundy, J.D. *et al.* (1985) Assessment of crash-worthy car materials. *Chartered Mechanical Engineer*, April, 31–5.

Guignard, J.C. (1971) Human sensitivity to sound and vibration. *Journal of Sound and Vibration*, **15**, 11–16.

Hagiwara, K. *et al.* (1983) A proposed method of predicting ship collision damage. *International Journal of Impact Engineering*, **1**, 257–80.

Harger, J.H. and Huelke, D.F. (1970) Femoral fracture produced by projectiles: the effect of mass and diameter on target damage. *Journal of Biomechanics*, **3**, 487–93.

Harris, J. (1976) The design and use of the TRRL side impact dummy. In 20th Stapp Conference.

Harris, R. and Barnard, L. (1968) Experiences of hot shortness in the forging of certain low-alloy steels. In *Deformation Under Hot Working Conditions*, The Iron and Steel Institute.

Hartbower, C.E. (1969) Materials sensitive to slow rates of straining. In *Impact Testing of Metals*, ASTM Special Technical Publication no. 466.

Hopkins, H.G. (1968) The method of characteristics. In *Engineering Plasticity* (eds J. Heyman and F.A. Leckie), Cambridge.

Hull, D. (1983) Axial crushing of fibre reinforced composite tubes. In *Structural Crashworthiness* (eds N. Jones and T. Wierzbicki), Butterworth.

Hull, H.C. and Allen, K.W. (1981) Design of an actively controlled snow ski binding release. *Journal of Biomechanical Engineering*, **5**, 138–45.

Humphreys, J.S. (1965) Plastic deformation of impulsively loaded straight clamped beams. *Journal of Applied Mechanics*, **32**, 7–10.

Hurty, W.C. and Rubinstein, M.F. (1964) *Dynamics of Structures*, Prentice-Hall.

Jacobson, L.S. and Ayre, R.S. (1958) *Engineering Vibrations*, McGraw-Hill.

Johnson, W. (1972) *Impact Strength of Materials*, Edward Arnold.

Johnson, W. (1973) An elementary analysis of an energy absorbing device: the rolling torus load limiter. *International Journal of Mechanical Sciences*, **15**, 357.

Johnson, W. and Mamalis, A.G. (1978) *Crashworthiness of Vehicles*, Mechanical Engineering Publications.

Jones, N. (1984) Scaling of inelastic structures loaded dynamically. In *Structural Impact and Crashworthiness*, Vol. 1 (ed. G.A. Davies), Elsevier.

King, A.I. (1972) Human tolerance limitations related to aircraft crashworthiness. In *Dynamic Response of Structures* (eds G. Herrmann and N. Perrone), Pergamon.

King, A.I. and Chou, C.C. (1976) Mathematical modelling, simulation and experimental testing of biomechanical crash system response. *Journal of Biomechanics*, **9**, 301–17.

Kinslow, R. (ed.) (1970) *High Velocity Impact Phenomena*, Academic Press.

Kolsky, H. (1963) *Stress Waves in Solids*, Dover.

Kornhauser, M. (1967) *Structural Effects of Impact*, Cleaver-Hume.

Krabiell, A. and Dahl, W. (1981) Influence of strain rate and temperature on the tensile and fracture properties of structural steels. In *Advances in Fracture Research*, Proceedings 5th International Conference on Fracture (ed. D. Francois), Pergamon.

Kramer, M. *et al.* (1973) Fracture mechanism of lower legs under impact load. In 17th Stapp Conference.

Lee, E.H. and Morrison, J.A. (1956) A comparison of the propagation of longitudinal waves in rods of viscoelastic materials. *Journal of Polymer Science*, **19**, 93–110.

Lindberg, H.E. (1965) Impact buckling of a thin bar. *Journal of Applied Mechanics*, June, 315–22.

Lister, R.D. and Wall, J. (1970) Determination of injury threshold levels of car occupants involved in road accidents. In *International Automobile Safety Conference Compendium*, SAE, pp. 818–33.

Macaulay, M.A. and Penoyre, S. (1980) The development of impact test procedures for legislation. In *Progress Towards Safer Passenger Cars in the United Kingdom*, Institution of Mechanical Engineers.

McClintock, F.A. and Argon, A.S. (eds) (1966) *Mechanical Behaviour of Materials*, Addison-Wesley.

Mather, B.S. (1968) Observations on the effects of static and impact loading on the human femur. *Journal of Biomechanics*, **1**, 331–5.

Meng, Q. *et al.* (1983) Axial crushing of square tubes. *International Journal of Mechanical Sciences*, **25**, 697–712.

Mentel, T.J. (1958) The plastic deformation due to impact of a cantilever beam with an attached tip mass. *Journal of Applied Mechanics*, December, 515–24.

Miner, L.H. (1984) Penetration resistance of aramid fibres and their composites. In *Structural Impact and Crashworthiness*, Vol. 2 (ed. J. Morton), Elsevier.

Murray, N.W. (1983) The static approach to plastic collapse and energy dissipation in some thin-walled steel structures. In *Structural Crashworthiness* (eds N. Jones and T. Wierzbicki), Butterworth.

Neathery, R.F. (1974) Analysis of chest impact response data and scaled performance recommendations. In 18th Stapp Conference.

Newman, J.A. (1975) The use of the head injury criterion (HIC) in protective headgear evaluation. In 19th Stapp Conference.

Norris, C.H. *et al.* (1959) *Structural Design for Dynamic Loads*, McGraw-Hill.

Onabe, H. *et al.* (1984) Blade containment of jet engines. In volume of late presentations at Imperial College Conference on Structural Impact and Crashworthiness.

Open University (1974) *Instrumentation Units 8/9/10 Transducers 2: Acceleration, Vibration, Velocity, Flow.*

Oxley, P.L.B. (1974) Rate effects in metal working processes. In *Mechanical Properties at High Rates of Strain*, Institute of Physics.

Radon, J.C. and Fitzpatrick, N.P. (1973) Deformation of PMMA at high rates of strain. In *Dynamic Crack Propagation* (ed. G.C. Sih), Noordhoff.

Reid, S.R. (1983) Laterally compressed tubes as impact energy absorbers. In *Structural Crashworthiness* (eds N. Jones and T. Wierzbicki), Butterworth.

Sadeghi, M.M. (1984) Design of heavy duty energy absorbers. In *Structural Impact and Crashworthiness*, Vol. 2 (ed. J. Morton), Elsevier.

Samuelides, E. and Frieze, P.A. (1983) Strip model simulation for low energy impacts on flat-plated structures. *International Journal of Mechanical Sciences*, **25**, 669–86.

Sarrailhe, S.R. (1984) Do tougher standards lead to better helmets. In International IRCOBI Conference on the Biomechanics of Impacts.

Scott, R.A. (1978) Linear elastic wave propagation: an annotated bibliography. In *Shock and Vibration Digest*, Naval Research Laboratory, Washington DC, Vol. 10, nos. 2 and 3.

Searle, J.A. and Brabin, E.J. (1970) *The Invertube*, Motor Industry Research Association Bulletin no. 2.

Searle, J.A. *et al.* (1971) *The Suppression of Noise in Crash Test Records*, Motor Industry Research Association Bulletin no. 1

Sensicle, A. (1968) *Introduction to Control Theory for Engineers*, Blackie.

Snowdon, J.C. (1968) *Vibration and Shock in Damped Mechanical Systems*, Wiley.

Snyder, R.G. (1970) Human impact tolerance. In *International Automobile Safety Conference Compendium*, SAE, pp. 712–55.

Spiegel, M.R. (1965) *Laplace Transforms*, Schaum Publishing.

States, J.D. (1969) The abbreviated and the comprehensive research injury scales. In 13th Stapp Conference.

Stronge, W.J. *et al.* (1983) Long stroke energy dissipation in splitting tubes. *International Journal of Mechanical Sciences*, **25**, 637–48.

Stüwe, H.P. (1968) Do metals recrystallise during hot working? In *Deformation Under Hot Working Conditions*, The Iron and Steel Institute.

Symonds, P.S. and Mentel, T.J. (1958) Impulsive loading of plastic beams with axial constraints. *Journal of the Mechanics and Physics of Solids*, **6**, 186–202.

Thomson, R.G. *et al.* (1984) Research at NASA on crash dynamics. In *Structural Impact and Crashworthiness*, Vol. 1 (ed. G.A. Davies), Elsevier.

Thwaite, R.M. Haythorn (1968) A more rational approach to strain hardening data. In *Engineering Plasticity* (eds J. Hayman and F.A. Leckie), Cambridge.

Ting, T.C.T. (1965) Large deformation of a rigid, ideally plastic cantilever beam. *Journal of Applied Mechanics*, June, 295–302.

Trade and Technical Press (1972) *Handbook of Noise and Vibration Control*, 2nd edn.

Versace, J. (1971) A review of the severity index. In 15th Stapp Conference.

Viano, D.C. (1977) Considerations for a femur injury criterion. In 21st Stapp Conference.

Vigness, I. *et al.* (1957) Effect of loading history upon the yield strength of a plain carbon steel. In *The Properties of Materials at High Rates of Strain*, Institution of Mechanical Engineers.

Von Gierke, H.E. (1964) Biodynamic response of the human body. *Applied Mechanics Reviews*, **17**, 951–8.

Wagenaar, H.W. (1968) Torsional ductility and strength of low-carbon steels at elevated temperatures. In *Deformation Under Hot Working Conditions*, The Iron and Steel Institute.

Wigley, D.A. (1971) *Mechanical Properties of Materials at Low Temperatures*, Plenum Press.

Wismans, J. *et al.* (1982) Madymo – a crash victim simulation computer program for biomechanical research and optimization of designs for impact injury prevention. AGARD Meeting, Koln, Germany, April.

Wittlin, G. (1983) Aircraft crash dynamics: modelling, verification and application. In *Structural Crashworthiness* (eds N. Jones and T. Wierzbicki), Butterworth.

Yamada, H. (1970) *Strength of Biological Materials*, Williams & Wilkins, Boston.

Zukas, J.A. *et al.* (1982) *Impact Dynamics*, Wiley.

Index